Fundamentals of Ground Engineering

Fundamentals of Ground Engineering

Editor

Sidhrath Saini

scitus
academics

Fundamentals of Ground Engineering

Edited by **Sidhrath Saini**

Printed in 2017

ISBN: 978-1-68117-000-8

Library of Congress Control Number: 2015931524

SCITUS Academics LLC,
616, Corporate Way, Suite 2, 4766,
Valley Cottage, NY 10989

www.scitusacademics.com

Notice

Contents

Preface

Ground Engineering is an understanding of geological structures, materials and processes, combined with the systematic application of investigative, scientific and mathematical techniques to produce practical solutions to ground related problems for the benefit of society.

This book, Fundamentals of Ground Engineering, summarizes the basic principles and theories of soil mechanics, the procedures for creating a geotechnical model, and the common analyses for slopes, foundations, and walls.

Ground Engineering specialises in the provision of high quality geotechnical and geoenvironmental ground investigation and associated professional services to the civil engineering, construction, housing, environmental management and financial sectors.

This book covers the basic theories and routine analyses of ground engineering, Fundamentals of Ground Engineering incorporates with a plenty of diagrams. This book is an ideal reference tool for students, practicing civil engineers-senior and junior-and by engineering geologists.

Editor

Chapter 1

Comparison of Classification Algorithms and Training Sample Sizes in Urban Land Classification with Landsat Thematic Mapper Imagery

Congcong Li [1], Jie Wang [2], Lei Wang [2], Luanyun Hu [3] and Peng Gong [2,3,4,5,*]

[1]State Key Laboratory of Remote Sensing Science, and College of Global Change and Earth System Science, Beijing Normal University, Beijing 100875, China

[2]State Key Laboratory of Remote Sensing Science, Institute of Remote Sensing and Digital Earth, Chinese Academy of Sciences, Beijing 100101, China

[3]Ministry of Education Key Laboratory for Earth System Modeling, Center for Earth System Science, Tsinghua University, Beijing 100084, China

[4]Department of Environmental Science, Policy, and Management, University of California, Berkeley, CA 94720-3114, USA

[5]Joint Center for Global Change Studies, Beijing 100875, China

ABSTRACT

Although a large number of new image classification algorithms have been developed, they are rarely tested with the same classification task. In this research, with the same Landsat Thematic Mapper (TM) data set and the same classification scheme over Guangzhou City, China, we tested two unsupervised and 13 supervised classification algorithms, including a number of machine learning algorithms that became popular in remote sensing during the past 20 years. Our analysis focused primarily on the spectral information provided by the TM data. We assessed all algorithms in a per-pixel classification decision experiment and all supervised algorithms in a segment-based experiment. We found that when sufficiently representative training samples were used, most algorithms performed reasonably well. Lack of training samples led to greater classification accuracy discrepancies than classification algorithms themselves. Some algorithms were more tolerable to insufficient (less representative) training samples than others. Many algorithms improved the overall accuracy marginally with per-segment decision making.

INTRODUCTION

Since the launch of the first land observation satellite, Earth Resource Technology Satellite (later changed to Landsat-1) in 1972, substantial improvements have been made in sensor technologies. The spatial resolution has increased over 100 times, from the 80 m of the Landsat-1 to 0.41 m of the GeoEye-1 (Orbview-5) satellite. The spectral sampling frequency has increased nearly 100 times, from a few spectral bands to a few hundred spectral bands. Classification of land cover and land use types has been one of the most widely adopted applications of satellite data. Although a large number of algorithms have been developed and applied to map land cover from satellite imagery, and new algorithm proposers have reported improvements in accuracies of their mapping experiments (see [1] for a substantial review on classification algorithms), it is difficult to find a systematic comparison on the performance of newly proposed algorithms. This is particularly true for machine learning algorithms, as many of them have been introduced to the field of remote sensing for less than 10 years. Instead, classifier performance comparison has only been limited to the comparison of a new algorithm with a conventional classifier like the maximum likelihood classifier [2–4], or the comparison among a small number of two to three new algorithms [5]. Through meta-analysis of a large number of published literatures on land cover and land use mapping, Wilkinson [6] found that accuracy improvement of land cover and land use mapping by new algorithms are hardly observable. However, this kind of analysis compares classification accuracies in different literature reporting applications over different study areas and/or with different types of satellite data.

As the number of machine learning algorithms increases, it is beneficial for the user community of machine learning algorithms to gain a better knowledge on the performances of each algorithm. In the field of remote sensing image classification, a more comprehensive comparison of major machine learning algorithms is needed. This must be done with the same land cover and land use classification scheme and the same satellite image. It is generally believed that final image classification results are dependent of a number of factors: classification scheme, image data available, training sample selection, pre-processing of the data including feature selection and extraction, classification algorithm, post processing techniques, test sample collection, and validation methods [7]. The purpose of this research is to compare performances of 15 classification algorithms when applied to the same set of Landsat Thematic Mapper (TM) image acquired over Guangzhou City, China, while keeping the other factors the same. The urban area of Guangzhou has been selected for this purpose as it includes relatively complex land cover and land use patterns that are suitable for classification algorithm comparison. In addition to applying the algorithms on a pixel-by-pixel basis, we also tested the algorithms on a per-segment basis to compare the effect of including the object-based image analysis as a preprocessing step in the image classification process.

STUDY SITE AND DATA

Our study site is located in the north of the Pearl River Delta (23°2′–23°25′N, 113°8′–113°35′E). As the capital city of Guangdong, Guangzhou is one of the fastest growing cities in China. It contains the core part of Guangzhou Municipality, and its rural-urban fringe (Figure 1). It can be divided into three regions: forest in the northeast, farmland in the northwest, and settlement in the south. As Guangzhou has been among the first group of cities that have undergone rapid development for over 20 years, it has been studied extensively for land use and land cover mapping and change detection (e.g., [8–10]).

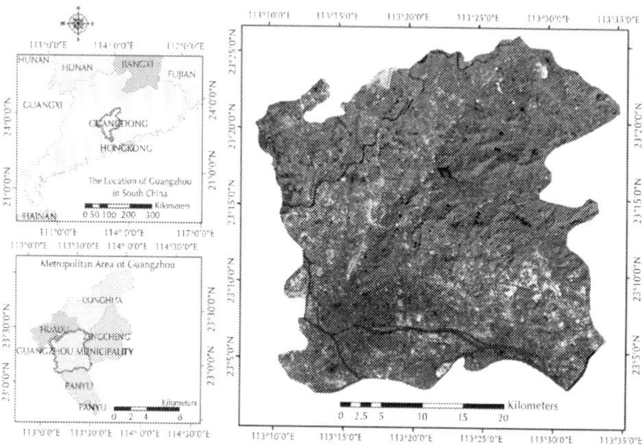

Figure 1: The study area. The image displays the green, red and near infrared band of the TM data with blue, green, and red color guns.

The Landsat Thematic Mapper (TM) image used here was acquired on 2 January 2009, in the dry season of this subtropical area. For classification on a single date image, there is no need to do atmospheric correction if the sky is clear, which is the case in this study [7,11]. Geometric correction was applied to the raw imagery by co-registering this image with a previously georeferenced TM image acquired in 2005. A total of 153 ground control points were selected from the image. A second order polynomial resulted in the root mean squared error of 0.44 pixels. The original image was radiometrically resampled with a cubic convolution algorithm (for classification purposes nearest neighbor or bilinear resampling would work as well). Due to its coarser resolution, we experimented with a 6-band set of the TM data by excluding the thermal band. In order to estimate the potential of satellite data of similar resolution but with only visible and near-infrared bands (e.g., the 32 m resolution multispectral camera on board the Disaster Monitoring Constellation satellites, the 30 m multispectral sensor on board China's Huanjing-1A satellite), we also experimented with a 4-band set of the TM data by further excluding the two middle infrared bands.

At the time of image acquisition, some fruit trees (such as Litchi) and several vegetables were in their blooming stage and some fruit trees (such as citrus) were in fruit-bearing stage. The elevation is high in the northeast mountains and low in the southwest farmlands. Newly developed industrial areas are in the southeast.

METHOD

Classification System

The land cover and land use classification system was developed to reflect the major land types in this area with reference to Gong and Howarth [2,12], and Gong *et al.* [13] Their meanings are self-explanatory (Table 1). On this basis, totally there were 14 subclasses for training, which were divided according to the spectral characteristics [2]. For example, the Industrial/commercial was subdivided into 4 types due to the spectral differences by different roofing materials.

Table 1: Land use classification system

Land-Use Types	Description
Water	Water bodies such as reservoirs, ponds and river
Residential area	Residential areas where driveways and roof tops dominate
Natural forest	Large area of trees
Orchard	Large area of fruit trees planted
Farmland	Fields where vegetables or crops grow
Industrial/commercial	Lands where roof tops of large buildings dominate
Cleared land/Land under construction	Lands where vegetation is denuded or where the construction is underway
Idle land	Lands where no vigorous vegetation grows

Training Samples

Training samples are primarily collected on a per-pixel basis to reduce redundancy and spatial-autocorrelation [7]. They were selected through image interpretation with intensive field visits over this area. Although more training samples are usually beneficial, as they tend to be more representative to the class population, a small number of training samples is obviously attractive for logistic reasons [14]. It is often recommended that a training sample size for each class should not be fewer than 10–30 times the number of bands [15–17]. This is usually okay for classifiers that require few parameters to be estimated like the maximum likelihood classifier when applied to a handful number of bands. With many classification algorithms, no previous study has reported an optimal number of training samples. To test the sensitivity of an algorithm to the size of training samples, we selected training samples uniformly from the images to make sure each subclass has 240 samples for later experiments. We sampled the training data to construct 12 sets of training samples with 20, 40, 60, 80, 100, 120, 140, 160, 180, 200, 220, and 240 pixels. For object-based method, we selected the segments contained the training pixels (240 pixels per subclass) as training objects.

Test Samples

We separately collected 500 pixels as test data, 138 of which were from field visits (done in April and December 2009, and June 2010), and the remaining were selected according to prior knowledge. The size of test sample for each land class was greater than 40 pixels (Figure 2). We used Kappa coefficient as the evaluation criterion [18].

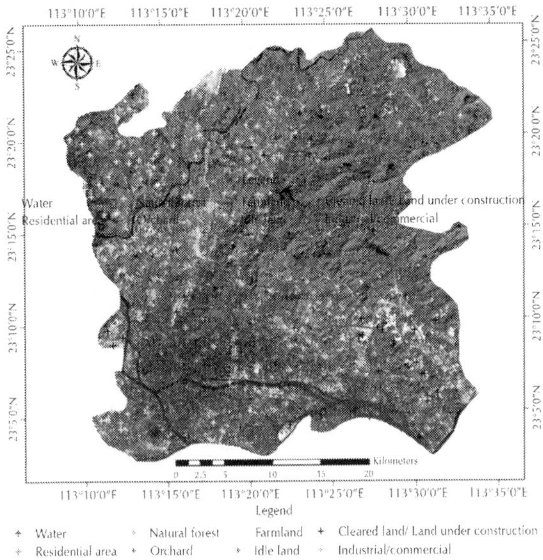

Figure 2: The distribution of test samples.

Classification Process

We tested 15 algorithms [19–33] all from easily accessible sources [34–37]. These algorithms are selected because they are openly accessible and easy to use. As the number of algorithms is large, and they are clearly documented elsewhere, sources of references on the codes and documentation of the algorithms are provided in Table 2. Most algorithms require certain parameterizations. While the choice of optimal parameter set is desirable, it is extremely difficult to do so even with the original algorithm developer as the application conditions vary so widely from one environment to another and from one data type to another. However, it is generally safe to adopt the recommended range by the algorithm developers. In practice this is usually what has been done. Therefore, we designed experiments to cover a majority of parameter combinations for each algorithm (Table 2) while adopting the parameter ranges as recommended in the original sources of references. These algorithms were tested using both pixel-based and segment-based methods. For the two unsupervised classifiers, the Iterative Self-Organizing Data Analysis Technique (ISODATA) is a popular advanced clustering algorithm [19] while the Clustering based on Eigenspace Transformation (CBEST) is an efficient k-means algorithm [20]. The clusters obtained were grouped into informational classes by the same analyst who did the selection of training and test samples. It is assumed that the analyst is most familiar with the study area given sufficient field visits and consulting with local experts.

Table 2: Algorithm parameter set up and source of codes

Algorithm	**Abbreviation**	**Parameter Type**	**Parameter Set**	**Source of Codes**
ISODATA	ISODATA	Number of Clusters Maximum Iterations	100, 150 100	ENVI
CBEST	CBEST	Number of Clusters Maximum Iterations	100, 150 100	CBEST
Maximum-likelihood classification	MLC	Mean and covariance matrix	Estimated from training samples	openCV
K-nearest neighbor	KNN	K weight	1,3,5,7,9,11 No weighting, 1/distance	Weka
Logistic regression	LR	Ridge estimator	0, 10^{-1}, 10^{-2}, 10^{-3}, 10^{-4}, 10^{-5}, 10^{-6}, 10^{-7}, 10^{-8}	Weka
C4.5	C4.5	MinNumObj Confidence	1, 2, 3, 4, 5, 6, 7, 8, 9, 10 0.05f, 0.1f, 0.2f, 0.3f, 0.4f, 0.5f	Weka
Classification and Regression Tree	CART	MinNumObj	1, 2, 3, 4, 5, 6, 7, 8, 9, 10	Weka
Quick, Unbiased, Efficient, and Statistical Tree algorithm	QUEST	Split types MinNumObj	univariate, linear 1, 2, 3, 4, 5, 6, 7, 8, 9, 10	QUEST

Random Forests	RF	numFeature numTrees	For 6-bands: 1,2,3,4,5,6 For 4-bands: 1,2,3,4 20, 40, 60, 80, 100, 120, 140, 160, 180, 200	Weka
Support Vector Machine	SVM	kernelType Cost gamma	radial basis function 10, 20, 30, 40, 50, 60, 70, 80, 90, 100 $2^{-2}, 2^{-1}, 1, 2^{1}, 2^{2}, 2^{3}, 2^{4}, 2^{5}, 2^{6}, 2^{7}$	Libsvm
Radial Basis Function Network	RBFN	MaxIteration NumCluster minStdDev	500, 1,000, 3,000, 5,000, 7,000 2, 3, 4, 5, 6, 7, 8, 9 0, 0.01, 0.05, 0.1	Weka
Logistic model tree	LMT	minNumInstances weightTrimBeta splitOnResiduals	5, 10, 15, 20, 25, 30 0, 0.01, 0.05, 0.1 False (C4.5 splitting criterion), True (LMT criterion)	Weka
Bagging C4.5	B_C4.5	bagSizePercent numIterations classifier	20, 40, 60, 80, 100 10, 50, 100, 150, 200 C4.5	Weka
AdaBoost C4.5	AB_C4.5	weightThreshold numIterations classifier	40, 60, 80, 100 10, 20, 30, 40, 50, 60, 70 C4.5	Weka
Stochastic gradient boosting	SGB	n.trees shrinkage bag.fraction	500, 1000 0.05, 0.1 0.2, 0.3, 0.4, 0.5, 0.6, 0.7, 0.8, 0.9	R

For segment extraction, we used BerkeleyImageSeg (http://berkenviro.com/berkeleyimgseg/) to perform image segmentations and then classified the segments by each algorithm. For the segmentation, the threshold is the most important parameter, which determines the size of the objects [38]. Here, four threshold values {5, 10, 15, 20} were examined. The shape parameter and compactness parameter were set to 0.2 and 0.7, respectively. The statistical spectral properties of the segments were then used in the segment-based classification. The features [39] are listed in Table 3. There are a total of 24 features. The parameters used for this method are selected according to the empirical values from the pixel-based classification, and taking the number of features used into consideration.

Table 3: The features used in the objected-based classification method

Feature
Maximum value of the segments for each spectral band (6 bands)
Mean or average values of the segments for each spectral band (6 bands)
Minimum values of the segments for each spectral band (6 bands)
The standard deviations of the pixels in the segments for each spectral band (6 bands)

Active Learning

Active learning is an algorithm for selecting effective training samples. This kind of algorithms adds unlabeled samples as training samples from the sample pool through human-machine interaction [40]. In this research, we used a margin-sampling algorithm [41,42], which takes advantage of SVM. It selected candidate samples lying within the margin of the model, and these samples are most conducive to the improvement of the classifier's performance. At the beginning, we randomly selected 20 samples for each class, and added 10 samples from the training set using margin sampling at a time. The best parameters of SVM are selected using simple grid search.

RESULTS

Pixel-Based Classification

Table 4 shows the best pixel-based classification accuracies of the algorithms. For the two unsupervised algorithms, they could produce as good results as some of the supervised algorithms when we cluster 150 spectral clusters. This is usually a very large number of clusters for an image analyst. Thus, we did not experiment for more clusters. Most supervised algorithms produce satisfactory results when the training samples are sufficient (more than 200 samples per class). However, MLC only requires 60 pixels to reach its highest accuracy. This indicates the high level of robustness and capability of generalization.

Table 4: Best classification accuracy for each algorithm using pixel-based approach

Algorithm	Parameter Choice	Accuracy
ISODATA	6-Band, Number of Clusters = 150	0.864
	4-Band, Number of Clusters = 150	0.841
CBEST	6-Band, Number of Clusters = 150	0.850
	4-Band, Number of Clusters = 150	0.846
MLC	6-Band, 60 training samples	0.892
	4-Band, 60 training samples	0.873
KNN	6-Band, 240 training samples, K = 3, weight = 1/distance	0.855
	4-Band, 240 training samples, K = 3, weight = 1/distance	0.823
LR	6-Band, 220 training samples, Ridge estimator = 10^{-8}	0.899
	4-Band, 200 training samples, Ridge estimator = 10^{-8}	0.862

C4.5	6-Band, 220 training samples, MinNumObj = 7, confidence = 0.2f	0.866
	4-Band, 240 training samples, MinNumObj = 3, confidence = 0.5f	0.841
CART	6-Band, 240 training samples, MinNumObj = 1	0.857
	4-Band, 220 training samples, MinNumObj = 2	0.818
QUEST	6-Band, 100 training samples, split type = linear, MinNumObj = 8/9/10	0.875
	4-Band, 180 training samples, split type = linear, MinNumObj = 1–10	0.843
RF	6-Band, 240 training samples, numFeatures = 1, numTrees = 20	0.873
	4-Band, 200 training samples, numFeatures = 1, numTrees = 60	0.848
SVM	6-Band, 240 training samples, kernelType = radial basis function, C = 80, gamma = 2^3	0.885
	4-Band, 240 training samples, kernelType = radial basis function, C = 50, gamma = 2^4	0.855
RBFN	6-Band, 220 training samples, minStdDev = 0.01, NumCluster = 9, MaxIts = 1,000	0.887
	4-Band, 240 training samples, minStdDev = 0.01, NumCluster = 8, MaxIts = 3,000/5,000	0.859
LMT	6-Band, 160 training samples, weightTrimBeta = 0, splitOnResiduals = C4.5 splitting criterion, minNumInstances ≥ 5	0.885
	4-Band, 200 training samples, weightTrimBeta = 0.1, splitOnResiduals = LMT splitting criterion, minNumInstances ≥ 5	0.862

B_C4.5	6-Band, 240 training samples, bagSizePercent = 80, numIterations = 100 C4.5parameter (MinNumObj = 2, confidence = 0.3f)	0.862
	4-Band, 240 training samples, bagSizePercent = 60, numIterations = 10 C4.5parameter (MinNumObj = 1, confidence = 0.2f)	0.836
AB_C4.5	6-Band, 220 training samples, weightThreshold = 40, numIterations ≥ 10, C4.5parameter (MinNumObj = 7, confidence = 0.2f)	0.866
	4-Band, 240 training samples, weightThreshold = 40, numIterations ≥ 10, C4.5parameter (MinNumObj = 5, confidence = 0.3f)	0.841
SGB	6-Band, 240 training samples, bag. fraction = 0.2, shrinkage = 0.1, n.tree = 1,000	0.869
	4-Band, 220 training samples, bag. fraction = 0.2, shrinkage = 0.1, n.tree = 1,000	0.852

A small value of K (K = 3) for KNN is the better choice in this study, and the distance-based weighting improves the KNN results. For the simple classification tree algorithms (CART, C4.5, and QUEST), minNumObj means minimum number of samples at a leaf node, which determines when to stop tree growing. All the three simple tree algorithms achieve high accuracies when this value is less than 10. In other words, they all grow big trees and then prune them. However, the LMT needs a large minNumInstances to build the tree. For RF, numFeatures means the number of features to be randomly selected at each node and numTrees means number of trees generated. Usually, the suggested value of numFeatures is $\sqrt{N}$, where N is the number of features [43]. However, in this research, we find a value smaller than $\sqrt{N}$ is more suitable. For SVM, we used radial basis function (RBF) kernel, the space affected by each support vector is reduced as the kernel parameter gamma increases. A slightly large gamma (2^3, 2^4) is the best choice for this research, which means more support vectors are used to divide the feature space. MinStdDev in RBFN is the minimum number of standard deviations for the clusters, controlling the width of Gaussian kernel function as gamma in SVM. numCluster is the number of clusters, determining the data centers of the hidden nodes. In this research, we found the numCluster equal to or slightly greater than the number of classes is a better choice. BagSizePercent in Bagging controls the percentage of training samples randomly sampled from the training sets with replacement. The results show that 60%–80% of the training set achieved better results. It is similar to weightThreshold in Adaboost, but the latter one resamples the training set according to the weight of the last iteration. It achieves good classification results using only 10 iterations. For SGB, bag.fraction controls the fraction of training set randomly selected without replacement. The best value of the sampling fraction is 0.2. This reduces the correlations between models at each iteration. The best shrinkage value, which is the learning rate is 0.1.

From Table 4 we can see that the best classification accuracy for the 6-band case is achieved by logistic regression, followed closely by the maximum likelihood classifier, neural network, support vector machine, and logistic model tree algorithms. Opposite to this, the CBEST and KNN produced the lowest accuracies. The range of Kappa coefficient from the lowest to the highest is 0.049. For the 4-band case, in general, there is a 0.02 to 0.04 difference in Kappa for each algorithm, confirming the fact that with fewer spectral bands there is indeed accuracy loss. However, in this experiment, the accuracy drop is quite small implying that the inclusion of the two middle infrared bands of the TM would not add a lot of power in separability to the classification of our classes. The maximum likelihood classifier produced the highest accuracy of 0.873 for the 4-band case, only 0.026 inferior to the highest accuracy with the 6-band case. The accuracy range for the 4-band case is between 0.818 and 0.873.

Objected-Oriented Classification

Table 5 shows the best classification accuracies using objected-oriented method. The results of this kind of classification are largely depending on the segmentation [44]. The classification accuracies are the highest when the segmentation scale is set to 5 (the smallest). The best performer is SGB with an accuracy improvement of 0.025 over the best pixel-based classification results. This is followed closely by RF. The accuracies decrease with the increase of the threshold. A higher threshold produces larger objects. For the TM image, which is 30 m in resolution, fragmentation is relatively high in this urban area. High threshold brings more mixed information in the segments under this classification system. As small segments are relatively homogeneous, the classifiers utilizing statistical properties of the segments rather than individual pixel values improved the results.

Table 5: Best classification accuracy for each algorithm using objected-oriented approach

	OB 5
MLC	0.898
KNN	0.871
LR	0.901
C4.5	0.912
CART	0.864
QUEST	0.882
RF	0.917
SVM	0.891
RBFN	0.894
LMT	0.908
B_C4.5	0.896
AB_C4.5	0.891
SGB	**0.924**

Comparing Table 5 with Table 4, we can see all results are improved based on objected-oriented approach using spectral features only. Among them, SGB produced the best results, followed by RF, C4.5, LMT, LR, and MLC. From another perspective, these algorithms could deal with high-dimensional data.

DISCUSSIONS

Most Common Errors among the Classifiers

Figure 3 shows the test pixels of different classes that have been misclassified at least once. The clusters of repeatedly misclassified pixels are mainly found in the urban areas. The red circled area is the center of Guangzhou, where many residential, forest, commercial buildings, and construction land are mixed. The blue circled area is the new urban district, where bare land and industrial park are mixed. Most of the algorithms perform poorly in these complex areas given the fact that it is easy to have a wider range of spectral characteristics than it would normally have in natural environment within the same class. In addition, in the green circled area, there are black greenhouses and iron and steel enterprise. They are misclassified as residential area or water. From Figure 4 we can see residential area and water; natural forest and orchard; industrial/commercial and cleared land/land under construction, residential, water; farmland and idle land are more easily mixed in the feature space.

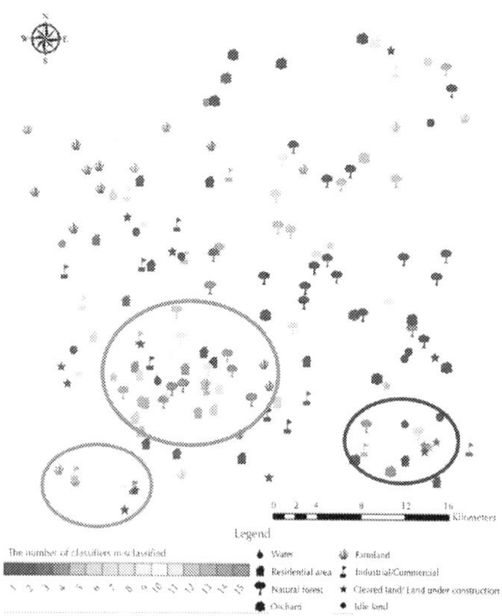

Figure 3: The distribution of misclassified test samples.

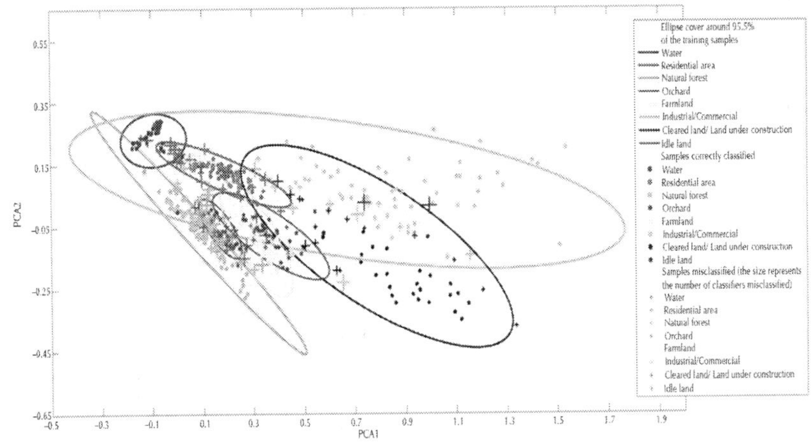

Figure 4: The distribution of the test samples in the feature space (principal component analysis (PCA) is used to reduce the dimension of data space).

Tables 6 and 7 show that residential area and water, and residential area and industrial/commercial are well distinguished by ISODATA, CBEST, MLC, and LR. RBFN is good at distinguishing forest from orchard, while SVM is good at classifying industrial/commercial and cleared land/land under construction. LR, SVM, and LMT can better distinguish farmland and idle land.

Table 6: The number of misclassified pixels by different algorithms

Class	Water	Residential Area	Natural Forest	Orchard	Farmland	Industrial/ Commercial	Cleared Land/ Land under Construction	Idle Land
ISODATA	1	4	16	7	10	6	14	1
CBEST	0	7	13	8	12	7	7	11
MLC	1	5	20	2	6	8	3	2
LR	0	7	20	0	5	7	4	1
KNN	0	11	30	2	6	8	5	2
C4.5	2	10	15	4	8	11	5	3
CART	0	11	27	2	8	8	3	3
QUEST	0	14	18	4	5	7	4	2
RF	0	14	17	4	8	7	3	2
SVM	0	13	22	1	5	6	2	1
RBFN	0	10	12	3	7	8	3	6
LMT	0	8	21	1	4	9	5	2
B_C4.5	1	11	26	2	8	7	3	2
AB_C4.5	2	10	15	4	8	11	5	3
SGB	0	12	21	2	8	9	4	1

Table 7: The confusion matrix of the best result

Class Water		Ground Truth (pixels)								
		Residential Area	Natural Forest	Orchard	Farmland	Industrial/ Commercial	Cleared Land/ Land under Construction	Idle land	Total	
Classification results	Water	41	1	0	0	0	0	0	0	42
	Residential area	0	84	0	0	2	2	0	0	88
	Natural forest	0	0	63	0	1	0	0	0	64
	Orchard	0	0	18	48	0	0	0	0	66
	Farmland	0	0	2	0	72	0	0	1	75
	Industrial/ commercial	0	6	0	0	0	64	4	0	74
	Cleared land/Land under construction	0	0	0	0	0	4	40	0	44
	Idle land	0	0	0	0	2	1	0	44	47
	Total	41	91	83	48	77	71	44	45	500

The Comparison of the Algorithms Using Pixel-Based Method

Figure 5 summarizes classification accuracies from all the parameter combinations listed in Table 2 and with different-sized training sets. The two unsupervised classifiers are not included as they were only tested with two parameter settings. We can see that all algorithms tested in this research could achieve high accuracies with sufficient training samples and proper parameters. MLC and Logistic regression have superior performances to other algorithms as their accuracy range is narrow and they can be easily set to produce a high accuracy. Another traditional algorithm—K-nearest neighbor does not get as high accuracies as these two. From the ranges of the boxes, MLC, LR, and LMT are the most stable algorithms among all the algorithms.

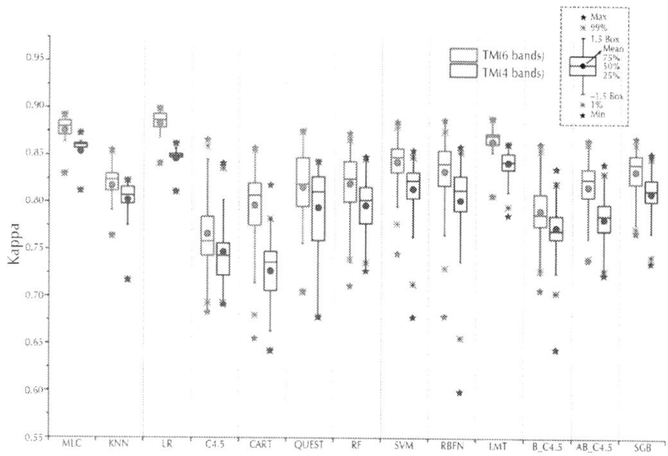

Figure 5: Comparison of the pixel-based supervised classification.

For the tree classifiers, the box ranges are large. They are sensitive to the selection of parameters and training samples. C4.5 and CART tested in this research both select only one feature to split on the nodes, while QUEST uses a linear combination of features to split classes. The latter divides the feature space more reasonably and flexibly when the spectral distribution is complex. RF as an advanced tree algorithm uses Bagging algorithm to generate different training sample sets, and ensembles the different trees created by these training sets. Shown in the figure, RF is superior to Bagging (C4.5) and other simple trees. Compared with Bagging, RF splits the node using features randomly selected. This can reduce the correlation between the trees, and then improve the stability of the classification results. SVM and RBFN show similar performance in our experiments, but the parameters of these two classifiers are difficult to set. In general, users could not get the most out of the two algorithms because of the difficulties in parameter setting.

The Adaboost shows better results than Bagging. It focuses on the wrongly classified samples in the previous iteration rather than randomly selected samples. Bagging and Adaboost classifiers are both built on different training sample sets. Their maximum accuracies are not higher than that of C4.5 indicating a larger variability of single tree classifiers like C4.5 but better stability with ensemble classifiers through Bagging and Boosting. SGB is another boosting algorithm, and it outperformed Adaboost. It fits an additive function to minimize the residuals at each iteration. It relies on the small data set randomly selected while Adaboost relies on the incorrectly classified samples. LMT is built on different classifiers. The algorithm is a tree classifier and builds logistic regression models at its leaves. It takes advantage of the decision tree, building logistic regression at a small and relatively pure space. The training sets have been divided into smaller subclasses according to the spectral characteristics. Therefore, LMT does not fully show its advantage.

The Impact of Different Training Set Sizes

Figure 6 shows the impact of training sample sizes on different classifiers. When the number of training samples is very small (e.g., 20, 40 samples), no algorithm performs well. The algorithms most affected by training sample size are the classification tree algorithms except for RF and Adaboost. They need sufficient samples to build the trees. MLC, LR, SVM, and LMT are the least affected algorithms. They could produce relative high accuracies using a smaller sized training set, and achieve stable results when there are 60 or more samples per class. All the algorithms except for MLC are improved in varying degrees by adding training samples. Generally speaking, MLC, LR, SVM, RBFN, and LMT could produce good results with small sized sample sets.

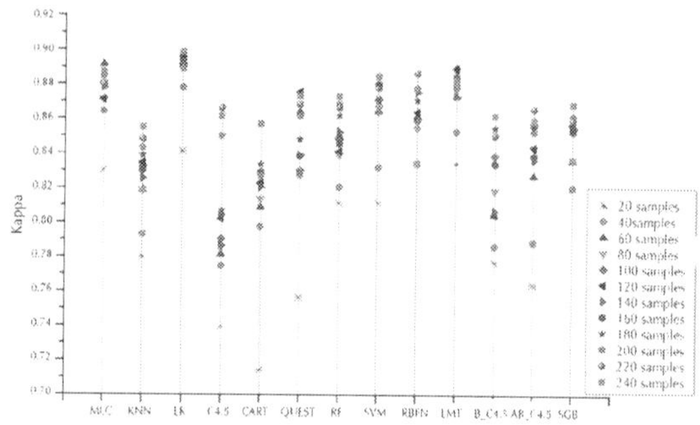

Figure 6: Pixel-based supervised classification with different sizes of training sets.

Active learning algorithm is used to test the representativeness of the training samples. Figure 7 shows that when the total number of training samples increases to 560–840 (40–60 samples for each subclass), about 1/6–1/4 of the entire training set, the classification results are satisfactory. The results are relatively stable when the training sample size further increases. In other words, the whole training set only contains 1/6–1/4 useful information. On the contrary, we randomly added the training samples without any rules, and then the results increased slowly and became stable when the training samples were representative enough (at about 2,800 samples, 200 samples for each subclass). That is why most of the algorithms achieve their highest accuracies when there are more than 200 samples per class.

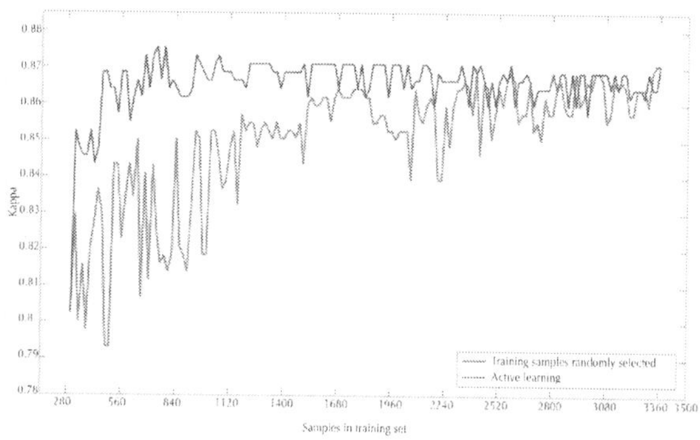

Figure 7: Active learning result and results based on training samples randomly selected.

Under such circumstances, using active learning algorithm to select training samples is an efficient way to achieving the optimal results before a large amount of trial and error tests. Active learning should be applied for representative sample selection to feed subsequent classification algorithms.

Algorithm Performances with Low-Quality Training Samples

The quality of training samples is reduced with the increase of thresholds in segmentation as segment statistics are becoming increasingly contaminated by information from potentially other classes. We could use this to assess algorithm performance. An algorithm showing good results on different thresholds performs well with low-quality training sets. Figure 8 shows that MLC, LR, RF, SVM, LMT, and SGB algorithms could deal with the contaminated information better than others. They could build more robust classifiers from weak training sets. In comparison, Adaboost algorithm shows good results only when the segmentation scale is smaller than 15. KNN results decrease sharply when the segmentation scale is greater than 10. As we have described hereinbefore, MLC, LR, SVM, RBFN, and LMT could perform well using a small training set. However, with deteriorating training samples, MLC, LR, SVM and LMT could still perform well.

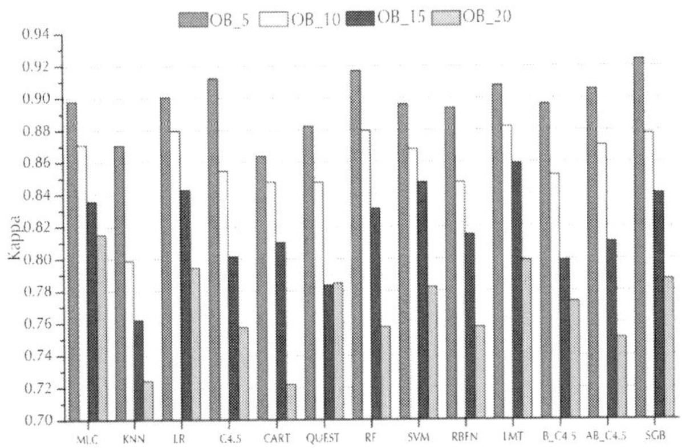

Figure 8: Objected-based supervised classification results.

SUMMARY

In this study, we compared 15 classification algorithms in the classification of the same Landsat TM image acquired over Guangzhou City, China, using the same classification scheme. The algorithms were tested on a pixel-based and segment-based classification. In the pixel-based decision making, the algorithms were tested with two band sets: a 4-band set including only the visible and near infrared TM bands and a 6-band set with all TM bands excluding the thermal and panchromatic bands. All supervised classifiers were tested with 12 sets of different sized training samples. In the segment-based decision making, the algorithms were tested with different segment sizes determined by different scale factors. All tests were evaluated by the same set of test samples with the total overall accuracy measured by the Kappa coefficient. The results can be summarized in the following:

1. The 4-band set of TM data by excluding the two middle infrared bands resulted in Kappa accuracies in the range between 0.818 and 0.873. The inclusion of the two middle infrared bands in the 6-band case increased this range to 0.850 and 0.899. This indicates the potential loss of overall accuracies in urban and rural urban fringe environments with the lack of middle infrared bands could be within 3%–5%.
2. Unsupervised algorithms could produce as good classification results as some of the supervised ones when a sufficient number of clusters are produced and clusters can be identified by an image analyst who is familiar with the study area. The accuracy of the unsupervised algorithms produced better than 0.841 Kappa accuracies for the eight land cover and land use classes.
3. Most supervised algorithms could produce high classification accuracies if the parameters are properly set and training samples are sufficiently representative. In this condition, MLC, LR, and LMT algorithms are more proper for users. These algorithms can be easily used with relatively more stable performances.
4. Insufficient (less representative) training caused large accuracy drops (0.06–0.15) in all supervised algorithms. Among all the algorithms tested, MLC, LR, SVM, and LMT are the least affected by the size of training sets. When using a small-sized training set, MLC, LR, SVM, RBFN, and LMT performed well.

5. In segment-based classification experiments, most algorithms performed better when the segment size was the smallest (with a scale factor of 5). At the scale of 5, SGB outperformed all other algorithms by producing the highest Kappa values of 0.924 and this is followed by RF. All algorithms are less sensitive to the large increase of data dimensionality. MLC, LR, RF, SVM, LMT, and SGB algorithms are the best choices to do the classification. They could produce relatively good accuracy at different scales.

With the increasing number of new algorithms emerging rapidly, there is a need to assess their performance and sensitivities to various kinds of environments. This need is best addressed by developing standard image sets with adequate classification scheme and sufficiently representative training and testing samples. This research represents one of such attempts. However, more datasets containing high quality training and test samples should be established for different types of remotely sensed data sets over typical environments in the world to support more objective assessment of new algorithms. Only when more comprehensive test data sets covering major environmental types of the world can we make more appropriate selection of algorithms for a particular application of remote sensing classification. Another important aspect that has not been assessed in this research is feature extraction and use of non-spectral features whose effectiveness has been demonstrated in the literature [45–51]. Furthermore, use of multisource data including optical, thermal and microwave data in urban land classification should be systematically evaluated [52,53]. Lastly, more analysis of the representativeness of training samples should be done in developing algorithm test image sample sets [54]. These will be evaluated in a future research.

This work was partially supported by the National High Technology Research and Development Program of China (2009AA12200101, 2013AA122804), the National Natural Science Foundation of China (41001275), the Dragon Program and a research grant from Tsinghua University (2012Z02287). Field work has been done with assistance by Zheng Lin and Xia Li at Sun Yat-Sen University.

REFERENCES

1. Lu, D.S.; Weng, Q.H. A survey of image classification methods and techniques for improving classification performance. *Int. J. Remote Sens* 2007, *28*, 823–870.
2. Gong, P.; Howarth, P.J. Land-use classification of SPOT HRV data using a cover-frequency method. *Int. J. Remote Sens* 1992, *13*, 1459–1471.
3. Pu, R.L.; Landry, S.; Yu, Q. Object-based detailed land-cover classification with high spatial resolution IKONOS imagery. *Int. J. Remote Sens* 2011, *32*, 3285–3308.
4. Shackelford, A.K.; Davis, C.H. A hierarchical fuzzy classification approach for high resolution multispectral data over urban areas. *IEEE Trans. Geosci. Remote Sens* 2003, *41*, 1920–1932.
5. Tzeng, Y.C.; Fan, K.T.; Chen, K.S. An adaptive thresholding multiple classifiers system for remote sensing image classification. *Photogramm. Eng. Remote Sens* 2009, *75*, 679–687.
6. Wilkinson, G.G. Results and implications of a study of fifteen years of satellite image classification experiments. *IEEE Trans. Geosci. Remote Sens* 2005, *43*, 433–440.
7. Gong, P.; Howarth, P.J. An assessment of some factors influencing multispectral land-cover classification. *Photogramm. Eng. Remote Sens* 1990, *56*, 597–603.
8. Fan, F.L.; Wang, Y.P.; Qiu, M.H.; Wang, Z.S. Evaluating the temporal and spatial urban expansion patterns of Guangzhou from 1979 to 2003 by remote sensing and GIS methods. *Int. J. Geogr. Inf. Sci* 2009, *23*, 1371–1388.
9. Fan, F.L.; Weng, Q.H.; Wang, Y.P. Land use and land cover change in Guangzhou, China, from 1998 to 2003, based on Landsat TM/ETM+ imagery. *Sensors* 2007, *7*, 1223–1342.
10. Seto, K.; Woodcock, C.; Song, C.H.; Huang, X.; Lu, J.; Kaufmann, R.K. Monitoring land-use change in the Pearl River Delta using Landsat TM. *Int. J. Remote Sens* 2002, *23*, 1985–2004.
11. Song, C.; Woodcock, C.E.; Seto, K.C.; Lenney, M.P; Macomber, S.A. Classification and change detection using Landsat TM data: When and how to correct atmospheric effects? *Remote Sens. Environ* 2001, *75*, 230–244.

12. Gong, P.; Marceau, D.; Howarth, P.J. A comparison of spatial feature extraction algorithms for land-use mapping with SPOT HRV data. *Remote Sens. Environ* 1992, *40*, 137–151.
13. Gong, P.; Wang, J.; Yu, L.; Zhao, Y.; Zhao, Y.; Liang, L.; Niu, Z.; Huang, X.; Fu, H.; Liu, S.; *et al.* Finer resolution observation and monitoring of global land cover: First mapping results with landsat TM and ETM+ data. *Int. J. Remote Sens* 2013, *34*, 2607–2654.
14. Foody, G.M.; Mathur, A.; Sanchez-Hernandez, C.; Boyd, D.S. Training set size requirements for the classification of a specific class. *Remote Sens. Environ* 2006, *104*, 1–14.
15. Mather, P.M. *Computer Processing of Remotely-Sensed Images*, 3rd ed; John Wiley & Sons, Ltd: Chichester, UK, 2004.
16. Piper, J. Variability and bias in experimentally measured classifier error rates. *Pattern Recognit. Lett* 1992, *13*, 685–692.
17. Van Niel, T.; McVicar, T.; Datt, B. On the relationship between training sample size and data dimensionality: Monte Carlo analysis of broadband multi-temporal classification. *Remote Sens. Environ* 2005, *98*, 468–480.
18. Congalton, R.G.; Green, K. *Assessing the Accuracy of Remotely Sensed Data: Principles and Practices*; CRC Press: Boca Raton, FL, USA, 1999; p. 137.
19. Ball, G.H.; Hall, D.J. *Isodata, a Novel Method of Data Analysis and Pattern Classification*; Stanford Research Institute: Menlo Park, CA, USA, 1965.
20. Chen, Y.; Gong, P. Clustering based on eigenspace transformation—CBEST for efficient classification. *ISPRS J. Photogramm. Remote Sens* 2013, *83*, 64–80.
21. Aha, D.W.; Kibler, D.; Albert, M.K. Instance-based learning algorithms. *Mach. Learn* 1991, *6*, 37–66.
22. Le Cessie, S.; van Houwelingen, J.C. Ridge estimators in logistic regression. *Appl. Stat* 1992, *41*, 191–201.
23. Quinlan, R. *C4.5: Programs for Machine Learning*; Morgan Kaufmann Publishers: San Mateo, CA, USA, 1993.
24. Breiman, L.; Friedman, J.H.; Olshen, R.A.; Stone, C.J. *Classification and Regression Trees*; Wadsworth Press: Monterey, CA, USA, 1984.

25. Loh, W.Y.; Shih, Y.S. Split selection methods for classification trees. *Stat. Sin* 1997, *7*, 815–840.
26. Breiman, L. Random forests. *Mach. Learn* 2001, *45*, 5–32.
27. Chang, C.C.; Lin, C.J. LIBSVM: A library for support vector machines. *ACM Trans. Intell. Syst. Technol* 2011, *2*, 1–27.
28. Bishop, C.M. *Neural Networks for Pattern Recognition*; Oxford University Press: New York, NY, USA, 1995.
29. Niels, L.; Mark, H.; Eibe, F. Logistic model trees. *Mach. Learn* 2005, *95*, 161–205.
30. Sumner, M.; Frank, E.; Hall, M.A. Speeding up Logistic Model Tree Induction. Proceedings of the 9th European Conference on Principles and Practice of Knowledge Discovery in Databases, Porto, Portugal, 3–7 October 2005; pp. 675–683.
31. Breiman, L. Bagging predictors. *Mach. Learn* 1996, *24*, 123–140.
32. Freund, Y.; Robert, E. Schapire: Experiments with a New Boosting Algorithm. Proceedings of the 13th International Conference on Machine Learning, Bari, Italy, 3–6 July 1996.
33. Friedman, J.H. Stochastic gradient boosting. *Comput. Stat. Data Anal* 2002, *38*, 367–378.
34. Bradski, G. The OpenCV library. *Dr. Dobb's J. Softw. Tools* 2000, *25*, 120–125.
35. Witten, I.H.; Frank, E. *Data Mining, Practical Machine Learning Tools and Techniques*, 2nd ed; Morgan Kaufmann Publishers: San Francisco, CA, USA, 2005; p. 525.
36. Hall, M.; Frank, E.; Holmes, G.; Pfahringer, B.; Reutemann, P.; Witten, I.H. The WEKA data mining software: An update. *SIGKDD Explor* 2009, *11*, 10–18.
37. Ridgeway, G. Generalized Boosted Models: A Guide to the GBM Package. Available online: https://r-forge.r-roject.org/scm/viewvc.php/*checkout*/pkg/inst/doc/gbm.pdf?revision=17&root=gbm&pathrev=18 (accessed on 21 September 2009).
38. Clinton, N.; Holt, A.; Scarborough, J.; Yan, L.; Gong, P. Accuracy assessment measures for object based image segmentation goodness. *Photogramm. Eng. Remote Sens* 2010, *76*, 289–299.
39. Clinton, N. BerkeleyImageSeg User's Guide. *CiteSeerX* 2010. Available online: http://citeseerx.ist.psu.edu/viewdoc/downloa

d?doi=10.1.1.173.1324&rep=rep1&type=pdf (accessed on 16 January 2014).

40. Tuia, D.; Volpi, M.; Copa, L.; Kanevski, M.; Munoz-Mari, J. A survey of active learning algorithms for supervised remote sensing image classification. *IEEE J. Sel. Top. Signal Process* 2011, *5*, 606–616.
41. Campbell, C.; Cristianini, N.; Smola, A.J. Query Learning with Large Margin Classifiers. In the Proceedings of the 17th International Conference on Machine Learning, Stanford, CA, USA, 29 June–2 July 2000.
42. Schohn, G.; Cohn, D. Less is More: Active Learning with Support Vector Machines. Proceedings of the 17th International Conference on Machine Learning, 29 June–2 July 2000; San Francisco, CA, USA.
43. Strobl, C.; Malley, J.; Tutz, G. Supplement to "An Introduction to Recursive Partitioning: Rational, Application, and Characteristics of Classification and Regression Trees, Bagging, and Random Forests". Available online: http://supp.apa.org/psycarticles/supplemental/met_14_4_323/met_14_4_323_supp.html (accessed on 11 November 2010).
44. Wang, L.; Sousa, W.; Gong, P. Integration of object-based and pixel-based classification for mangrove mapping with IKONOS imagery. *Int. J. Remote Sens* 2004, *25*, 5655–5668.
45. Gong, P.; Howarth, P.J. The use of structural information for improving land-cover classification accuracies at the rural-urban fringe. *Photogramm. Eng. Remote Sens* 1990, *56*, 67–73.
46. Gong, P.; Howarth, P.J. Frequency-based contextual classification and grey-level vector reduction for land-use identification. *Photogramm. Eng. Remote Sens* 1992, *58*, 423–437.
47. Xu, B.; Gong, P.; Seto, E.; Spear, R. Comparison of different gray-level reduction schemes for a revised texture spectrum method for land-use classification using IKONOS imagery. *Photogramm. Eng. Remote Sens* 2003, *6*, 529–536.
48. Liu, D.; Kelly, M.; Gong, P. A spatial-temporal approach to monitoring forest disease spread using multi-temporal high spatial resolution imagery. *Remote Sens. Environ* 2006, *101*, 167–180.

49. Cai, S.S.; Liu, D.S. A comparison of object-based and contextual pixel-based classifications using high and medium spatial resolution images. *Remote Sens. Lett* 2013, *4*, 998–1007.
50. Wolf, N. Object features for pixel-based classification of urban areas comparing different machine learning algorithms. *Photogramm. Fernerkund* 2013, *3*, 149–161.
51. Luo, L.; Mountrakis, G. Converting local spectral and spatial information from a priori classifiers into contextual knowledge for impervious surface classification. *ISPRS J. Photogramm. Remote Sens* 2011, *66*, 579–587.
52. Weng, Q.H. Remote sensing of impervious surfaces in the urban areas: Requirements, methods, and trends. *Remote Sens. Environ* 2012, *117*, 34–49.
53. Gamba, P. Human settlements: A global challenge for EO data processing and interpretation. *IEEE Proc* 2013, *101*, 570–581.
54. Mountrakis, G.; Xi, B. Assessing reference dataset representativeness through confidence metrics based on information density. *ISPRS J. Photogramm. Remote Sens* 2011, *78*, 129–147.

Chapter 2

The Legal Structure of Taiwan's Wetland Conservation Act

Yi-Yuan Su [1,2]

[1]Graduate School of Law, Hokkaido University, Sapporo 060-0809, Japan

[2]Department of Law, National Chung Hsing University, Taichung 402, Taiwan

ABSTRACT

In July of 2013, Taiwan passed its Wetland Conservation Act and will begin the implementation of the Act on 2 February 2015. With this Act, Taiwan has become the second Asian country to have specific legislation on wetland conservation and protection. This new law enables the society to achieve sustainable utilization on wetland ecological services. The core concepts of the Wetland Conversation Act include biological diversity conservation and wise use of wetland resources. Special political circumstances prevent Taiwan from registering its wetlands as a conservation priority under the Ramsar Convention. This new law allows the government to evaluate and assign a specific area as a "Wetland of Importance." Under this status, any development activities within the designated area shall be prohibited unless the developer prepares a usage plan for review. The usage plan and the original usage of the natural resources within the wetland area shall also follow the "wise use" principle to protect the wetland and biological service system. However, this new law does not provide clear separation between the two different "wise use" standards. If the development is deemed necessary, new law provides compensation mitigation measures to extend the surface of the wetland and provides additional habitats for various species. Wetland conservation and management rely heavily on systematic research and fundamental data regarding Taiwan's wetlands. Determining how to adopt these scientific methodologies and transfer them into enforceable mechanisms is a sizeable challenge for both biologists and lawyers as the Wetland Conservation Act creates many legal norms without clarifying definitions. This article will review the current wetland regulations from the legal perspective and provide suggestions for enforcement in the future.

INTRODUCTION

The Legislative Yuan of the Republic of China (known as Congress of Taiwan) passed its Wetland Conservation Act (hereinafter referred to as the Wetland Act) contains 42 articles [1] and was decreed by Presidential Order No. Hua-Zong-Yi-Yi 10200127201 on 3 July 2013 [2]. The implementation date was set for 2 July 2014, but was changed to 2 February 2015 by the Executive Order Yuan-Tai-Chian-Tze No. 1030030131 on 12 June 2014 [3]. The Wetland Act is a unique piece of legislation in Northeast Asia. Only South Korea and Taiwan have specific laws to protect their domestic wetlands. The Wetland Act is a voluntary implementation measure put forth by the non-Ramsar Convention Party (The Convention on Wetlands of International

Importance Especially Waterfowl Habitat, Ramsar, 1971; hereinafter referred to as the Ramsar Convention) [4].

Taiwan's establishment of the Wetland Act arose from protesting of a development project of the Kuo-Kuang petrochemical plant project in 2005. The site of the development project was located on the delta of the Zuo-Shui River in central Taiwan. The delta is an intertidal zone with abundant marine species and is the habitat of the endangered migratory bird, the Eastern Curlew (Numenius madagascariensis). [5] Both citizens and environmental groups challenged the necessity of building petrochemical factories in this important wetland. They expressed their deep concern that the development of the petrochemical industry in this area would negatively impact the local environment and public health in the future. During six years of discussion and debating, some environmental groups and citizens implored the government to create a legislative proposal to protect wetlands on the island from further rapid, economic developments. They insisted the definition of wetland cover various types of wetlands, but not be limited to coastal area wetlands. Accordingly, the Ministry of the Interior, the party responsible for the legislation, assigned the Coastal Restoration Urban and Rural Development Branch of the Construction and Planning Agency to study and draft a proposal for wetland protection in 2009. At the same time, some environmentalists, scholars, and NGO members merged their opinions to prepare their own proposals for further negotiation with government. In the final stages, there were five different proposals raised to the Legislative Yuan by various representatives. In April of 2012, during the proposal negotiation and editing, the President cancelled the petrochemical plant project because of adverse environmental impact concerns from the project and the rise of the global crude oil price. However, the wetland protection legislation process continued and finally passed on 14 June 2013.

The wetland legislation in Taiwan is a great achievement of sustainable utilization on Taiwan's wetland ecological services; especially because it was contributed from a button up proposal from citizens. This Wetland Act not only provide an effective measure for the country to protect its wetland resources but also prove the influences to an non-Ramsar Contracting Patty from the international treaties. The international treaties do provide proper references materials to the individual county. They enhance the global wetland protection through promoting domestic legislation and enforcements.

GENERAL LEGAL PRINCIPLES

Article 18 of the Basic Environment Act is the source of the legal authority supporting the Wetland Conservation Act [6]. The Basic Environmental Act provided "sustainable development" and "polluter pay" principles to guide all environment regulations in Taiwan. Article 18 states all levels of government entities shall take aggressive measures to protect wetland environments. The central government responded to Article 18's requirement to establish wetland protection regulations and followed its general principles to design regulations for wetland conservation. However, there are eight chapters in the Wetland Act, and none of them adopt the "sustainable development" and "polluter pay" principle [5] described in the Basic Environment Act. On the other hand, Article 1 of the Wetland Act states the core concepts of this Wetland Act are "biological diversity conservation" and "wise use" of wetland ecological services [1]. This means the measurements and regulations are based on "wise use" as defined in Article 4 [1].

Because the Basic Environment Act is the basic and fundamental regulation for all environmental regulations in Taiwan, it is not clear whether the "wise use" principle used in the Wetland Act fits the sustainable development principle described in the Basic Environment Act. Paragraph 2, Article 2 of the Basic Environment Act defines sustainable development as "satisfying contemporary needs without sacrificing the ability of future generations to satisfy their needs" [6]. This definition follows the definition provided by the World Commission on Environment and Development in 1987. However, the purpose and definition of the "wise use" principle is obviously different from the sustainable development definition as the sustainable development, or "wise use" definition, is more focused on the sustainable utilization of wetland resources only. It is not clear whether the usage of the wetland ecological services will only concern the stable conditions of wetlands and ignore the sustainable development of economic, society and institutions, particularly the biological diversity conversation goal of this new bill.

Article 4 of the Wetland Act states the wise use principal is predominantly focused on the sustainable utilization on biological resources, water resources, and land resources within its ecological capacities and achieves stable conditions of wetland resources [2]. As part of environment regulations, this new law does not follow and connect with the regulations provided by the Basic Environment Act.

The Ministry of Interior explained the principle of "wise use" is a legal standard adopted from the decision made by the third meeting of Conference of the Contracting Parties of Ramsar Convention in 1987 [7]. However, Taiwan is not a member of the Contracting Party of the Convention because of its political status in the United Nations. The Taiwanese government introduced a legal standard directly from this international treaty without considering its own domestic legal systems and authorizations. Taiwan voluntarily implemented the principle of the Convention, but did not consider that the fundamental requirement of this principle requires various methodologies and scientific data regarding wetland resources. Generating such scientific data and conducting the necessary investigation requires time, research, and funding and are imperative in making decisions regarding wise use of wetland resources. The lack of fundamental, scientific data on wetland resources leads to failed implementation and enforcement. Proper preparation of the scientific data necessary for the wetlands in island can take over one year. The major concern for the implementation of this principle has been modified and improved by the Conference of the Contracting Parties after 1987. The decision made by the conference has updated the conceptual framework and guidelines of the "wise use" principle. Because Taiwan is not a Contracting Party to the Ramsar Convention, it cannot participate in the real time negotiation and update of the wise use principle. Instead, Taiwan must go through the legislative process. Additional time is needed to update and change its legal definitions and negotiation within various levels. If Taiwan could join the negotiation of the Ramsar Convention, the Administrative Branch could more easily adopt a new definition of the wise use principle. The government restricted its definition in 1987 on the Wetland Act, and the Judicial Branch cannot use an updated concept on any related issue. The extension of the implementation date of the Wetland Act to February 2015 is a useful opportunity for the government to establish the scientific data and database for wetlands.

FRAMEWORK AND STRUCTURE

Based on the principles of biological diversity conservation and wise use, the Wetland Act created four conceptual structures to design its eight chapters that will manage wetland resources in Taiwan. The biological diversity conservation principle provides basic concepts to evaluate and delimit the wetland of importance, in addition to providing the guideline for wetland conservation and usage plan evaluation, review, and identification. The usage and management of wetland resources shall follow the "wise use" principle. This principle also applies to development avoidance, impact mitigation, and ecological compensation. The government also established a new wetland fund and label system for conservation management. The penalty articles are provided in the last chapter of the act.

WISE USE PRINCIPLE

The first chapter of the Wetland Act is the general principle chapter comprised of seven articles, including the purpose of this new Act. There are multiple authorities in wetland governance. The Ministry of Interior is the assigned authority at the central government level; the municipal governments and city/county governments are the authorities at the local level. Different authorities have different missions. Article 4 provides the definitions of the terms used in this act and clarify the difference between ecological compensation and habitat compensation. Article 5 defines the boundaries of the wise use principle and conservation of both wetland nature resources and ecological function. Article 5 confused the definition in Article 4, subparagraph 4 with the definition of "wise use". According to Article 5, "wise use" requires all levels of government to adopt adequate management and wise use of natural resources and ecological functions for sound, stable ecological system management on a "no net loss" base. Additionally, wise use requires every level of government to promote sustainable development of the entire environment by enhancing wetland conservation and restoration. This article further defines the principle of conservation and wise use in its subparagraphs as

i. "Natural wetlands shall be protected with priority, and the water resource system also maintained.

ii. Enhance the conservation of the wetland's plant and animal resources.

iii. Wetlands offering ecological network significance and its peripheral environment and landscape shall be adequately and comprehensively planned and maintained.

iv. In coordination with wetland restoration, flood prevention, flood retention, water quality purification, water resource conservation and utilization, landscape and recreation, an overall planning shall be promoted for the wetland system. If necessary, constructed wetlands may be created in proper areas by suitable means [2]."

Subparagraphs I and II of Article 5 discuss issues on restoration and conservation, but not the usage of wetland resources. Subparagraphs III and IV mention conservation planning, but not use of wetland resources. Comparing Article 5 with the terms used in Article 4, "wise use" is defined as:

"Referring to a timely, fittingly, moderately and adequately sustainable utilization of the biological resources, water resources and land resources that is achieved in a compatible approach within the wetland ecological capacity to maintain the quality and quantity of wetland resources in a stable condition"[2].

This Article mentioned sustainable utilization of wetland resources. In reviewing the terms used in Article 5, one finds a greater focus on conservation and restoration, but not wise use. Having two different definitions of wise use in the same Wetland Act is a source of some confusion. Article 4 is designed to define terms and norms used in this new law, while Article 5 provides general principles on wetland conservation and management. Obviously, Article 5 has more influence on related measures and policies on wetland resource management and usage. Although Article 4 provides a general definition of wise use, it defines the legal norms only, but does not create a guideline or standard function on the related measures and policies. The general definition of norms and terms cannot replace the general principle function. Such a discrepancy may adversely affect the implementation of the "wise use" principle or lead to a wrongful interpretation of the conservation principle that would involve the elements of "wise use" principle.

The Ministry of Interior explained the wise use principle used in this new Act was taken from the Resolution of Ramsar Convention in

1987. Based on the Annex of Recommendation 3.3 of the 3rd Meeting of the Conference of the Parties of Convention on Wetlands in 1987, "wise use" is defined as the following:

"The wise use of wetlands is their sustainable utilization for the benefit of humankind in a way compatible with the maintenance of the natural properties of the ecosystem" [7].

In this international legal document, the elements of the wise use principle are "sustainable utilization" and "natural properties of the ecosystem." [7,8,9] The definition in Article 4 of the Wetland Act did not adopt the terms of "natural properties of the ecosystem," but used the terms "ecological capacities" and "biological resources" in the text. The seventh Meeting of Ramsar COP adopted "ecological character" in 1999. [10] It is obvious that Taiwan's Wetland Act neither adopts the "nature properties of the ecosystem" definition made in the 1987 Recommendation nor the "ecological character" definitions made in the 1997 Ramsar Resolutions. However, the text of Article 4 is more similar to the wording in the 1987 Recommendation document because they both mentioned water, soil, and other natural resources. In Taiwan's Wetland Act, the use of natural resources is different from the usage of ecological resources in the Wetland Act. There is no further definition of the term of ecological capacities in Article 4 or the text of other articles. Whether the term ecological resources is equal to the Ramsar Convention is not clear and may be left to the judicial authority for further explanation. However, the later Ramsar Resolution updated the concept of the wise use after 1987. Since the Taiwan government insisted on the adoption of the "wise use" definition made by the Regina Meeting in 1987, the court may be limited to its use and will not be able to accept the new definition included in the ecological characteristic terms after 1990 and 1996. [11] Therefore, the implementation and enforcement on the wise use on the wetlands of this island is worth further observation.

WETLANDS OF IMPORTANCE

Article 8 of the Wetland Act refers to the wetlands identified by this Act as Wetlands of Importance. A Wetland of Importance is classified in three levels: the international level, national level, and regional level. The competent central government authority uses eight standards to evaluate and identify a Wetland of Importance, including biodiversity of the wetland, natural elements, habitat of migratory species, presence of rare and endangered organisms, water and soil, historical culture, folklore heritage, landscape aesthetics, and ecological function. A Wetland of Importance can also be amended or abolished when wetland status is changed by natural disaster or is faced by the need of public interest. In comparing this with the conservation principle described in Article 5, determining how to evaluate the need of public interest and conservation of a Wetland of Importance became a challenge on the government.

A public hearing and participation for any evaluation and abolishment of Wetland of Importance are necessary and required by Article 10. Article 10 also requires public notice for an evaluation that shall be announced and posted to public media thirty days before the evaluation meeting is hosted. The competent central government authority shall consider all suggestions and opinions from the general public. The government shall also prepare proper responses to the public's related opinions and publish them in the public media. These regulations adopted democratic procedures and expanded the opportunity for public participation. These regulations for public participation not only disclose related information on wetland resources, but also increase the awareness for wetland protection.

Thus far, the central government authority has not established procedures for evaluation and abolishment meetings. The lack of established procedures is important as the government begins reorganizing from its 2013 structure. As part of this reorganization, the central government authority over wetland protection may be shifted from the Ministry of Interior to the "Ministry of Natural Resources and Environmental Protection" [12]. It is unclear whether the government can establish the required public hearing procedure before the implementation date.

UTILIZATION PLAN ON WETLAND OF IMPORTANCE

Article 13 requires the competent central government authority to produce the National Wetland Conservation Guideline every five years to promote wetland conservation strategies and mechanisms. A Utilization Plan on Wetlands of Importance is also required. They shall be proposed and prepared by the competent government authorities in three levels of Wetland of Importance. The Utilization Plan on Wetland Conservation shall not only follow the conservation principle, but shall also state twelve subject matters described in Article 15 of the Wetland Act [13]. In order to confirm the water supply of a Wetland of Importance, the fifth paragraph of Article 15 requires the central government authority to jointly define the standards of water resources within the Wetland of Importance with related government agencies, which included definitions of irrigation, drainage, reservoir, supply, and underground water.

The Wetland of Importance shall be divided into five areas based on ecological function. These areas include core conservation area, ecological restoration area, environmental education area, service area, and others. The development and construction in core conservation areas and ecological restoration of international and national Wetland of Importance is strictly prohibited.

The Utilization Plan for the Wetland Conservation Act shall be prepared and formulated within one year after the evaluation announcement is published to the general public and media. The publication and evaluation procedures on the Utilization Plan on Wetland Conservation are similar to the procedure of the Wetland of Importance evaluation. The utilization plans shall be reviewed every five years after their announcement and implementation.

WISE USE ON WETLANDS OF IMPORTANCE

Based on the conservation principle, natural wetlands will be protected as a priority. When the government is preparing a national park, regional plan, or urban plan, Article 20 requires the government to consult the central government authority if the border of the plans overlap with a Wetland of Importance. The reviewing or developing of a hydrological plan and soil conservation plan will be subject to consultation by the central government authority to prevent possible influence on a Wetland of Importance. Those who implement the environmental impact assessment shall also consult the central government authority if the development project could potentially impact the Wetland of Importance. When other agencies are reviewing or planning a development project, they shall consult with the central government authority as well.

The Wetland Act recognizes any building or land used for agriculture, fisheries, or salt-making industries located within the border of a Wetland of Importance as a continuing condition. If the use of the land violates any other regulation, the other regulation shall be adopted to address those issues. The Wetland Act identified three industries (agriculture, fisheries, and salt industries) that are permitted to operate within a border of a Wetland of Importance. However, land usage and regional plans in Taiwan are always mixed with various industries. Whether the other industries located within the parameter of a Wetland of Importance could be eliminated depends on the enforcement of all levels of governments after the implementation date. Additional efforts to investigate every activity in the scope of the Wetland of Importance are necessary for establishing a date for decision-making, but these efforts will consume time and financial resources. Furthermore, there are concerns regarding the stakeholders near a Wetland of Importance because Article 20 permits only three industry activities (agriculture, fisheries, or salt-making industries) within the border and determines methods to deal with various stakeholders, and beneficiaries of the interest on the wetlands might be a challenge.

If the utilization of the continuing conditions causes significant impact on the Wetland of Importance, the government authorities shall order the developer, managing entity, or user to improve their utilization measures by a prescribed date. If necessary, the government authorities may provide consultation to improve the utilization measures to a standard of "wise use." The term "wise use" is defined in Article 4. If the utilization is overdue or the improvement failed to fit wise use and caused net loss in a Wetland of Importance, the violators will be penalized and the government will implement impact mitigation, offset compensation, and ecological compensation as described in Article 27.

The owner of private land located on a Wetland of Importance shall not increase facilities or extend the facility's surface area. The owner must obtain a permit from the government authorities for any changes made to the land.

For any privately owned land located within the border of a Wetland of Importance, the government authorities can adopt proper regulations to expropriate, allocate, or lease the land for a conservation utilization proposal when necessary. Any public land within the border of a Wetland of Importance may be commissioned to private operator with permission from the government authority. Those operators who are earning profits and running travel services shall apply for a permit from the government authority and pay a specific percentage of the profit as rebates. Other prohibited activities are described in Article 25. These prohibited activities include emitting any pollutant, hunting, and drawing water resources.

DEVELOPMENT AVOIDANCE, IMPACT MITIGATION, AND ECOLOGICAL COMPENSATION

Although natural wetlands are a prioritized protected resource, Article 27 allows for the development and utilization on wetlands, except for in those wetlands designated as a Wetland of Importance. The development and utilization on the wetland shall be reviewed with a Wetland Impact Report and approved by government authorities. The reviewing principle mandates following the orders of avoidance, mitigation, offset, and compensation described in the first paragraph of Article 27. The offset and compensation measures and standards will be produced and announced by the government authorities.

This article's text is fundamentally similar to the Resolution XI. 9 adopted by the 11th Meeting of the Conference of the Parties to the Ramsar Convention [14]. The issue here is the definition of "wise use" in Taiwan's Wetland Act. The act used the term "ecological characteristic" in its text, but explained that the idea was developed from the 1987 definition and used the term of "ecological capacities" in the text. The concept of avoidance, mitigation, offset, and compensation was not developed in 1987. Before the government creates the measures and standards for offset and compensation, it must consider whether it could adopt the standards suggested by the Ramsar Resolution in 2012. This is a concern because Taiwan is not bound by the Convention and its 1987 "wise use" definition. The central government authority shall not only create these standards and measures by making Enforcement Rules for this Act, but will also address the interpretation of the term ecological capacity. The Annex on the 2012 Ramsar Convention was established and based on the core concept of ecological characteristic, but not ecological capacity, which is used by Taiwan's Wetland Act. The basic concept of offset and compensation provides different reviewing standards and measures on wetland protection. The government shall consider this fundamental discrepancy and carefully implement explanations of these different terms. Therefore, joint cooperation and research between the science and social science profession is necessary and essential. If they fail to do so and adopted the suggested guidelines from the international documents, it will lead to many legal concerns and various conflicts on choosing the proper standard. Because the Enforcement Rules of the Wetland Act have not been established, the government still has a limited amount time to work on these critical interpretations on different terms of ecological capacities and ecological characteristics.

OTHERS

The Wetland Act also established the Wetland Fund for the implementation needs of wetlands conservation measures and policies. The financial sources for the Wetland Fund come from government finances, rebates, interests from the fund management, and other sources. The usage of the fund is limited to specific purposes described in Article 34, including promotion of the awareness on wetland conservation and protection, research and monitoring, education, and others.

In order to encourage public participation on wetland conservation and promote wetland environmental education, the central government authority shall establish wetland label and pay specific rebates to the Wetland Fund. The management measures, policies, and audition rules on wetland label system shall be established and announced by the competent central government authority.

CONCLUSIONS

A wetland is a complex system and its management measures rely on dependable scientific data for better conservation. The new Wetland Act passed in a very short time, and now, only has a one-year implementation period. So far, most of the policy surrounding Taiwan's wetlands are lacking scientific detail and biological data for further conservation planning. However, the Taiwanese government has identified over eighty-four Wetlands of Importance on the island [15]. Identifying the wetlands became a concern because the border of those wetlands may overlap with some private land, and related land acquisition or leasing rules or procedures are not yet established. The government may face various challenges from private landowners regarding border allocation. Without further preparation and scientific data to explain and negotiate with neighboring stakeholders, it will be difficult to promote public participation in wetland conservation and protection. The definition of "wise use" is also a great concern in the ongoing rule-making because its definition was self-created, but does not follow the definition given by the Ramsar Convention. However, the government claimed the Ramsar Convention is its major legal resource for future policy-making, citing the term "ecology capacities" used by the Wetland Act. The government will need to further clarify and defined the difference between "ecological characteristic" and "ecological capacity" in order to determine other guidelines related to "wise use" of Taiwan's wetlands. Taiwan's Wetland Conservation Act will be implemented in 2015. We expect the wetland resources in Taiwan to be well protected by strong enforcement and competent implementation of the regulations. The legal community will be conducting further research and paying close attention to the interpretation of the legal standards in this new law as it moves forward.

ACKNOWLEDGMENTS

The author would like to give his special thanks to the Ministry of Science and Technology Republic of China (Taiwan) for the continuing support on this research (MOST 101-2627-B-005-005).

REFERENCES AND NOTES

1. The Wetland Conservation Act. *The Legislative Yuan Gazette*, 2013.
2. The Wetland Conservation Act. *President Office Gazette*, 2013.
3. The Wetland Conservation Act. *The Executive Yuan Gazette*, 2014.
4. Convention on Wetlands of International Importance Especially as Waterfowl Habitat. 1971.
5. International Union for Conservation of Nature (IUCN). The IUCN Red List of Threatened Species, February 2014. Available online: http://iucnredlist.org (accessed on 22 August 2014).
6. The Basic Environment Act. *President Office Gazette*, 2002.
7. Wise use of wetland: Annex, Definition of wise use, Recommendation 3.3, Ramsar Convention: Regina, Canada, 27 May–5 June 1987. Available online: http://ramsar.rgis.ch/cda/en/ramsar-documents-recom-information-on-wise-use/main/ramsar/1-31-110^23129_4000_0__ (accessed on 10 December 2014).
8. Annex to the Convention on Wetlands, Recommendation 3.3, Ramsar Convention: Regina, Canada, 27 May–5 June 1987. Available online: http://ramsar.rgis.ch/cda/en/ramsar-documents-recom-information-on-wise-use/main/ramsar/1-31-110^23129_4000_0__ (accessed on 10 December 2014).
9. Bowman, M.J. The Ramsar Convention Comes of Age. *Netherlands Int. Law Rev.* 1995, *42*, 1–52.
10. Pritchard, D. *Wise Use of Wetlands: Concepts and Approaches for the Wise Use of Wetlands*, 4th ed.; Ramsar Handbooks for the Wise Use of Wetlands. Ramsar Convention Secretariat: Gland, Switzerland, 2010; p. 6.
11. Additional Guidance for the Implementation of the Wise Use Concept, Resolution 5.6, Ramsar Convention: Kushiro, Japan, 9-16 June 1993; Guidelines for the Implementation of the Wise Use Concept, Recommendation 4.10, Ramsar Convention: Montreux, Switzerland, 27 June–4 July 1990. Available online: http://ramsar.rgis.ch/cda/en/ramsar-documents-resol-additional-

guidance-for/main/ramsar/1-31-107^20915_4000_0__ (accessed on 10 December 2014).

12. Amendment on Executive Yuan Organization Act. *President Office Gazette*, 2010.

13. The twelve matters described in Article 15 are "I. he border and year period of the plan; II. The guidance of the superior and relevant programs; III. The investigation and analysis of local society and economy. IV. The rudimentary investigation and analysis of water resource system, ecological resources and environment. V. Land and building utilization status. VI. Primarily protected areas with critical scientific search, cultural heritage and ecological and environmental value. VII. Stipulations or measures governing the wetland system functional zoning, and its conservation, restoration, restricted or banned conducts, and maintenance. VIII. Permissible wise use items and management. IX. Water resource protection and utilization management. X. Emergency response and recovery. XI. Financial and implementation. XII. Other pertinent matters."

14. Annex, An Integrated Framework and Guidelines for Avoiding, Mitigating and Compensating for Wetland Losses, Resolution XI.9, Ramsar Convention: Bucharest, Romania, 6–13 July 2012. Available online: http://ramsar.rgis.ch/pdf/cop11/res/cop11-res09-e.pdf (accessed on 10 December 2014).

15. Map of Taiwan's Wetlands, Construction and Planning Agency Ministry of Interior. Available online: http://wetland-tw.tcd.gov.tw/WetLandWeb/web-en/map-wetlands.html (accessed on 26 August 2014).

Chapter 3

On the Design and Response of Domestic Ground-Source Heat Pumps in the UK

Chris Underwood

Faculty of Engineering & Environment, Northumbria University, Newcastle NE1 8ST, UK

ABSTRACT

The design and response of ground source heat pumps coupled to vertical closed loop arrays in UK domestic applications are investigated in this article. Two typical UK house types are selected as the vehicle for the study and a detailed dynamic thermal modelling method is used to arrive at time-series heating demands for the two houses. A new empirical heat pump model is derived using experimental data taking into account the deteriorating performance of the heat pump during periods of light load. The heat pump model is incorporated into an existing numerical ground model and completed with a classical effectiveness type heat exchange model of the closed loop array. The model is used to analyse array sizing and performance over an extended time period, as well as sensitivity of the design to soil conductivity and borehole heat exchanger resistance and sensitivity to over-sizing and part-load behavior of the heat pump. Results show that the UK's standard for ground source design (the Microgeneration Certification Scheme) may lead to under-estimated array sizes and that heating system over-sizing and deleterious part-load heat pump performance can add up to 20% to the electrical consumption of these systems.

INTRODUCTION

During the first phase of recent UK field trials on 81 domestic air- and ground-source heat pumps the median seasonal performance factor (*SPF*) of the sample of 54 ground source heat pumps was found to be 2.2 [1]. (In this context, the seasonal performance factor is taken to mean the heat delivered to the space heating and domestic hot water service over a complete annual operating cycle divided by the corresponding electricity required by the heat pump and its associated source and sink circulating pumps.) The performance was well below expectations and was attributed to a multitude of factors including system sizing, type of source, building efficiency, user behavior and installation practices. Following the first phase of these trials, several site interventions were planned and improvements to the Microgeneration Certification Scheme (MCS) [2] were made to inform future users of heat pumps on procedures for design and installation. The interventions ranged from minor measures (e.g., adjustments to controls) to major interventions (in some cases, involving a re-installation of the heat pump or radiators where they were considered inappropriately sized). Following the interventions, a sample of 32 improved installations were monitored for one further year of which 21 were ground source heat pumps. The median *SPF* for this sample was found to be 3.1 [3]. Whilst the improvement in seasonal coefficient of performance to 3.1 is welcome, it is still short of the potential for these systems. Most of the domestic heat pumps available at present use a remarkably similar kit of parts. These include brazed plate heat exchangers for the evaporator and condenser (which tend to give better heat exchange "pinch" than the first-generation shell and tube heat exchangers), scroll fixed-speed compressors and mechanical ("thermostatic") expansion devices. A buffer water store on the heating side is usually included either as part of the heat pump package or plumbed-in separately and the control of the heat pump is usually by means of a heating thermostat mounted in the heating flow from the heat pump (though often in the heating return especially when a buffer tank is not used). Therefore, it is to be expected that variations of the kind identified in [1,2] are most likely to be due to design, installation and operational issues rather than issues concerned with the heat pump itself.

The UK experiences a quite different climate than that experienced in continental regions in that its weather is governed by maritime conditions with mild winters and frequent episodes of abrupt variations in weather from day to day. Partly because of this, lower standards of thermal insulation tend to be used in UK building construction than is the case in colder climatic regions. It is also generally accepted that the UK lags behind continental Europe and America in its exploitation of ground source heat (though the recent introduction of tariff incentives is likely to see that situation change quite abruptly in the coming years). The UK's geological formations (many of which are at least moderately water-bearing) are also quite different to many of the well-drained sites in America where closed loop ground source heat pumps have been widely used and reported. To account for this, reference will be made in this work to a new and large data set of ground formation conductivities from a large number of sites across the UK. Thus, UK conditions for the use of ground source heating offer a number of unique features which make detailed research and evaluation particularly timely.

In this work, the impact of heat pump capacity and ground array design for domestic ground source heat pumps operating in UK conditions are explored in detail with a view to establishing design criteria that might lead to improvements in operating performance. This will be achieved through the following objectives:

- Detailed dynamic modelling of two typical UK house types of differing sizes and energy demands.

- Development of a new empirical heat pump model which accounts for the degradation in part-load operating performance evident in these systems.
- Design of vertical ground loop arrays using the simulated house energy demands and the new heat pump model.
- Investigations into the sensitivity of the array designs to variations in soil and borehole heat exchanger properties and an analysis of the robustness of the designs over an extended operating time horizon (5 years).
- Investigations into the sensitivity of the seasonal heat pump electrical energy use and house comfort conditions to the under-sizing and over-sizing of the heating system.

Bagdanavicius and Jenkins [4] modelled electrical energy demands for a community of 96 two, three and four-bedroom houses with the assumption that all of the houses used a ground source heat pump sized according to the MCS heat pump standard [2]. They found domestic hot water energy use to have a major bearing on results but their results were restricted to just one winter week. In an experimental study, Blanco *et al.* report on the performance of a variable speed compressor-driven heat pump at domestic scale [5]. Variable speed heat pumps are yet to be widely adopted at domestic scale and are likely to bring part-load performance improvements as well as dispensing with the need for sink-side buffer heat storage. They found that, on average, the electrical consumption when operating with heating temperatures of 35 °C was 29% lower than when operating at 45 °C. Wood *et al.* carried out an experimental investigation into a vertical ground array consisting of single "U-tubes" cast into 10 m-deep piles [6]. They used 21 such piles making an overall array size of 210 m and demonstrated a typical heating delivery rate of 6 kW with a seasonal performance factor of 3.62. So called "energy piles" can provide strong economic advantages compared with conventional vertical ground source arrays though this method of foundation construction is seldom used in house construction. Boait *et al.* investigated the performance of a group of bungalows equipped with ground source heat pumps [7]. They obtained results that were consistent with the first phase of the Energy Saving Trust's trials mentioned earlier [1] and, again, revealed performances that were below observations from other field trials carried out elsewhere in continental Europe. This conclusion is also comprehensively arrived at by a detailed analysis of a number of European field trials on both air- and ground-source heat pumps carried out by Gleeson and Lowe [8]. One of a number of reasons for the performance shortfall mentioned in the Boait *et al.* study was the limited availability of very low capacity heat pumps for small well-insulated UK dwellings (such as bungalows) meaning that the larger capacity heat pumps that are used tend to operate for long periods at light load. However the overarching finding from all of these sources is that the reasons for the disappointing UK performances are complex and multi-faceted and further work is needed.

In conclusion, evidence is beginning to emerge which suggests that domestic ground source heat pumps in the UK are performing below expectations and below comparable installations in other parts of Europe. It appears that there are many reasons for this but a key consideration would appear to be the capacity of the heat pumps used in the UK in relation to the pattern of domestic energy demand. Matters are complicated by the requirement to generate domestic hot water through the summer months when space heating is usually not required. This is often at temperatures that are higher than would be required for space heating due to the need to ensure safe hygiene standards in hot water storage and plumbing systems (though this can be conveniently addressed with minimal loss in performance through the use of a two-zone de-superheater/condenser [5,9]).

ENERGY DEMAND MODELLING IN A SAMPLE OF UK HOUSES

House Type Selection and Seasonal Modelling

Nearly 55% of the 2.8 million UK houses surveyed in 2010 and reported in the National Energy Efficiency Data Framework (NEED) [10] had gross floor areas of between 51 m^2 and 100 m^2 and the next largest group (29%) had gross floor areas of between 101 m^2 and 150 m^2. The mean gas consumptions of these two groups in 2010 were 13,200 kWh and 18,000 kWh, respectively, which, with a well-maintained gas boiler efficiency of 0.85, can be considered to translate to thermal energy demands of 11,220 kWh and 15,300 kWh, respectively. Two house types with gross floor areas of 75 m^2 and 125 m^2 were therefore chosen to fall within this range. It was assumed that the first of these would be a two-story mid-terrace house and the second would be a two-story detached house. Simplified seasonal energy demand modelling was carried out using BREDEM 2012 [11].

Briefly, BREDEM 2012 [11] "Building Research Establishment Domestic Energy Model" is a calculation procedure to estimate all forms of energy consumption in houses. It is a monthly calculation method based on monthly-averaged weather data. The house to be modelled is split into two zones—a main living zone (usually the living room or main reception room) and the rest of the house forms a second balance zone. A simple time-constant-based model is used to estimate heating energy use whereas regression-fitted models based on observations of a large number of UK house types are used to calculate hot water and electrical equipment energy demands. Though the electrical demands are of no direct interest in the present work, they are indirectly used to provide information on internal heat gains which are used to adjust space heating demands. Mainly because of the regression models for energy uses that would otherwise be very difficult to calculate theoretically, BREDEM 2012 [12] tends to give highly representative and accurate results over longer-term averaging periods in strictly UK conditions but is not suitable for energy studies over short time periods. The methods described in BREDEM 2012 [11] form the primary methods used in all UK domestic energy planning and design evaluation work.

Adjustments were made to house layout, glazing and construction details such that results from the BREDEM modelling provided values that were similar to the two mean gas consumptions from the NEED data [10] with the intention of arriving at two "very typical" UK house types as judged by energy demand performance. The details arrived at in this way are consistent with houses that were either constructed or fully refurbished to standards prevailing at about the beginning of this century.

Thermal properties and other details of the two house types can be found in Appendix A (Table A1 and Table A2).

Results of the BREDEM modelling using a London site selection for both houses are summarized in Table 1. The totals given in Table 1 can be seen to be of the order of the NEED results mentioned above and can thus be considered to be representative of commonly occurring UK house types of this scale.

Table 1: Simulated annual energy demand for two typical UK house types

House Type	Space Heating (kWh)	Domestic Hot Water (kWh)	Total (kWh)
Mid-terrace	7,764	2,983	10,747
Detached	13,556	3,377	16,933

Dynamic Thermal Modelling

The BREDEM results give a good representation of annual energy demand in UK housing which is why the methods used are linked to the Standard Assessment Procedure (SAP) for the purposes of energy efficiency policy development, design and planning applications in the UK. The method is, however, limited in that for the detailed analysis of renewable, low carbon and other complex microgenerators, more granular time-series results of energy demand data are required (e.g., half-hourly or hourly data) whereas BREDEM is only able to provide results at monthly (minimum) intervals. For loads that can be considered to follow a daily average pattern such as domestic hot water this is of no consequence but for the dominant energy demand due to space heating which varies due to both climate and user activity this is a serious limitation. Indeed the need to achieve a better understanding of the response of the heat pump and heating system with the dynamics of the building were emphasised by Boait *et al.* in their field trial study [7]. The two house types were thus remodeled using a dynamic thermal modelling procedure and the results compared for accuracy with the BREDEM reference results.

Re-modelling of the space heating energy use for both houses was carried out using a superset of bespoke building energy modeling components developed for Simulink in the Matlab environment—the Simulink HVAC (heating, ventilating and air conditioning) Blockset. Details of the blockset library can be found in Appendix B (Figure B1 and Figure B2) and details of the mathematical derivations of the most relevant blocks used in the present work can be found in the literature [12,13]. All dimensional and other details as were used in the BREDEM modelling were also used in the dynamic thermal modelling (Appendix A). A current test reference year weather file for London was used from the Chartered Institution of Building Services Engineers (CIBSE) Future Weather Years set [14]. (Note that monthly-average values from this weather file were also used in the earlier BREDEM modelling.)

For domestic hot water energy, the original BREDEM monthly predictions were broken down into daily average values. The operating daily schedules were then applied which consisted of one 2 h heating period each weekday morning following by one 7 h heating period during each weekday evening. On weekend days, one single 16 h heating period was applied. The domestic hot water loads were allocated to the morning and evening periods in the ratio of 1:2, respectively (this pattern was assumed to apply on both weekdays and weekend days).

Annual total energy demands predicted using the dynamic thermal model were found to be 10,529 kWh for the mid-terrace house and 16,994 kWh for the detached house which agree very favourably with the totals from the reference model (BREDEM) given in Table 1. The annual distributions of these demands generated using the Simulink dynamic thermal model are plotted (including hot water demands) in Figure 1.

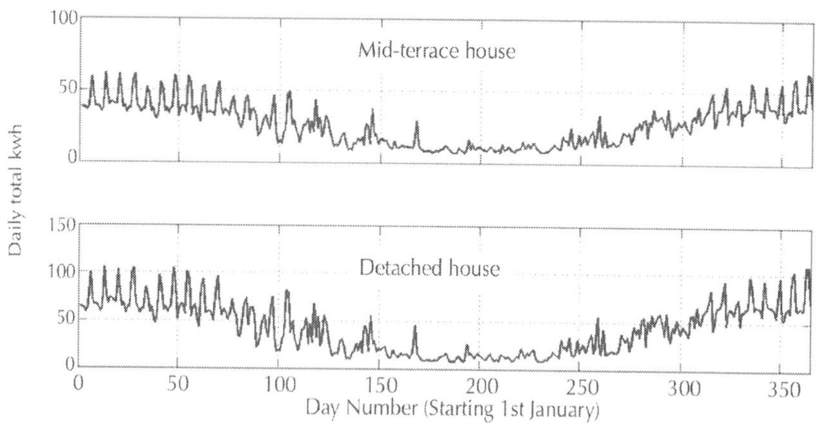

Figure 1: Simulated distributions of energy demand due to heating and hot water.

DEVELOPMENT OF A GROUND-SOURCE HEAT PUMP MODEL

One of the stated objectives of this work is to take account of the decline in heat pump performance when operating at part load. Conventionally, a heat pump model capable of describing part load performance would require being fully dynamic and, consequently, very computationally demanding. A simpler model is needed particularly when the heat pump merely forms part of a more extensive modelling problem (*i.e.*, treatment of the ground array as discussed in the next section). Historically, a simple (and conditionally accurate) modelling approach often used is the so-called "catalogue fit" model. In this, one or more dependent variables (such as electricity consumption and coefficient of performance) are fitted to a manufacturer's performance data set using multiple-regression, e.g., [15,16,17]. These models are accurate (at least as far as that particular manufacturer's product is concerned) but are limited in that the various standards from which these data are prepared assume static operation at the declared boundary conditions specified in the catalogue. In other words, they assume that the heat pump operates at full continuous capacity at the stated conditions. In practice, heat pumps like other energy generating plants will operate according to some control conditions and will spend large parts of the operating cycle at part load. When operating intermittently to meet a varying load under a thermostat, the first few seconds or minutes of operation as the heat pump starts is merely recovering losses prior to raising the heating temperature to a point where it can contribute to the prevailing load. Thus, in this initial phase of operation, the energy delivered by the heat pump constitutes a loss. This loss accumulates according to the number of thermostat starts the heat pump makes over a full operating cycle. Since most domestic heat pumps in use today for space heating are controlled by thermostats, losses can be significant and clearly become amplified when the heat pump capacity is larger than it needs to be (*i.e.*, when it is over-sized). To capture this behaviour whilst also retaining a simple model structure, an empirical model is developed based on a domestic-scale laboratory pilot.

Description of the Heat Pump Test Rig

The heat pump test rig consists of a conventional domestic scale water-to-water heat pump using refrigerant R410A with a rated (nominal) heating capacity of 9.5 kW. The heat pump was controlled using a thermostat (with an adjustable set-point) mounted in the heating system return connection at the inlet to the heat pump. Detailed physical information about the heat pump can be found in Appendix C (Table C1).

The heat pump is sourced from three 100 m (deep) plus one 65 m (deep) vertical closed loop borehole heat exchangers giving a total array capacity of 365 m. All four heat exchangers are connected in parallel and each can be individually isolated to enable a variable capacity array. Each borehole heat exchanger comprises a 32 mm high density polyethylene U-tube inside a 130 mm diameter borehole with the inner spaces grouted using thermally-enhanced bentonite. The array lies in Coal Measures and a thermal response test carried out shortly after installation gave a mean soil conductivity of 2.45 $W \cdot m^{-1} \cdot K^{-1}$ and a borehole thermal resistance of 0.162 $m \cdot K \cdot W^{-1}$. The undisturbed soil temperature was 12.7 °C. The source fluid is water-ethylene glycol mixture (10% ethylene glycol by volume).

The heat pump outputs to four equally-sized double panel convector-radiators which provide heating to the local laboratory environment. Each has a rated emission of 1.07 kW with reference to a mean water temperature of 40 °C and a local air temperature of 20 °C. Though the heat pump does not have a buffer tank, the heating system has a significant amount of "natural" buffering due to a low-loss header and significant runs of larger diameter (32 mm) steel piping upstream of the heating system. It is estimated that, collectively, these features provide approximately 50 L of system-side buffering that would not normally be present in a domestic installation.

Instrumentation consisted of a current clamp and voltage transducer on the incoming electrical connections to the heat pump. The power factor was measured separately using an electric circuit analyser and found to have an average value of 0.924 with very minimal variance. Source and sink heats were measured using a pair of resistance-wire temperature detectors and time-of-flight ultrasonic flow meters on both sides of the heat pump. Compound measurement uncertainties were assessed to be on average ±0.8 kW on heating loads (typically <10% of measured heat) and ±0.12 kW on active electricity use (typically <5% of measured electricity use). The laboratory air temperature was measured using several K-type thermocouples and subsequently averaged at each reporting time row.

Experimental Procedure

Forty-eight heat pump capacity tests were carried out using the following variations in plant configuration:

- 100 m, 200 m, 300 m and 365 m of source array capacity.
- 1, 2, 3 and 4 convector-radiators turned on.
- Nominal heating thermostat set point temperatures of 38 °C, 40 °C and 50 °C.

All variables were monitored at 5 s intervals and averaged in sets of six for reporting at 30 s intervals. The first and final rows of data during each on-phase were discarded in order to remove data spikes arising from incompleteness in the data during a reporting interval in which the thermostat's status changed. Individual borehole tests were carried out on separate days to allow for soil stabilization between tests and the first thermostat cycle of results between convector-radiator adjustment events were discarded to allow for heating system stabilization. Note that the test heat pump in this case is monovalent and the logged electrical consumption includes the source pump but not the sink (heating system) pump. Thus, when used for seasonal performance evaluations, the results based on the present work would lead to a seasonal performance factor that has become referred to in some of the literature as "SPF_{H2}" [3,8].

Results and Model-Fitting

It is well established that the performance of a heat pump depends *inter alia* on the source and sink temperatures. In addition, the part-load ratio (*P*) of the heat pump is defined here as the ratio of actual heating delivered to the maximum heating delivered. This can be determined from the results in one of two ways. Either by calculating the total heating delivered over time (in kWh) and dividing by the continuous average heating that would have been delivered over the same time period if the heat pump had not been operating intermittently; or by calculating the radiator emission from the measured heating water temperatures and laboratory air temperature. (Both methods were used and differences between them were found to be minor.)

To identify the influence of the array capacity and P on the heat pump performance, each set of test results was averaged and the heat pump coefficient of performance plotted against P for each discrete array capacity. Results, given in Figure 2, show a strong influence of both of these variables on performance. It is particularly noted that the heat pump performance falls sharply at P values of ≤0.5. Therefore, the fitted model was selected to account for variations in source and sink temperature, array capacity and P. For convenience, the source and sink temperature were reduced to a temperature differential, ΔT_{sosi}, between heat pump outlet sink water temperature and heat pump outflow source fluid temperature. Thus, for individual fitting over each array capacity, there will be two dependent variables; P and ΔT_{sosi}.

Two alternative model forms were tested; bi-linear (Equation (1)) and bi-quadratic (Equation (2)):

$$CoP = (a_1 + b_1 \times P) \times (c_1 + d_1 \times \Delta T_{sosi}) \quad (1)$$

$$CoP = (a_2 + b_2 \times P + c_2 \times P^2) \times (d_2 + e_2 \times \Delta T_{sosi} + f_2 \times \Delta T_{sosi}^{\ 2}) \quad (2)$$

(in which $a_1 ... d_1$ and $a_2 ... f_2$ are regression constants).

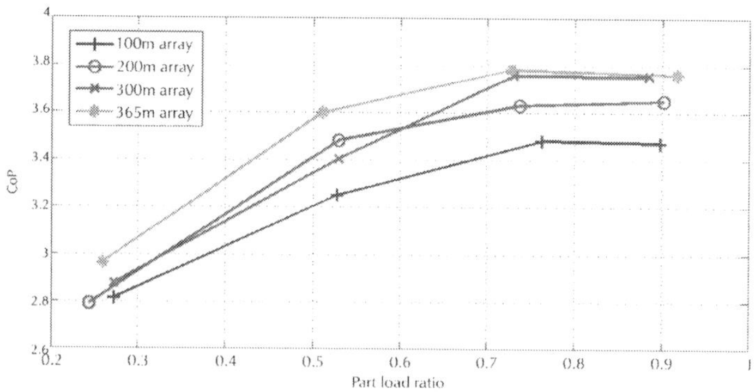

Figure 2: Part load performance results from the heat pump test rig.

Results of multiple-regression fitting over all viable data using both models are summarized in Figure 3 and Figure 4. For the bi-linear model, most of the target/model error values across all data are within ±10%. There is no improvement in goodness-of-fit when adopting the slightly more complex bi-quadratic model and the target/model error values here are mostly contained within the higher error range of ±20%. Hence, the bi-linear model is adopted in this work. Multiplying out Equation (1) gives the following for which values of the fitted constants, *A*, *B*, *C*, and *D*, for all array capacities investigated can be found in Table 2:

$$CoP = A + B \times P + C \times \Delta T_{sosi} + D \times P \times \Delta T_{sosi} \quad (3)$$

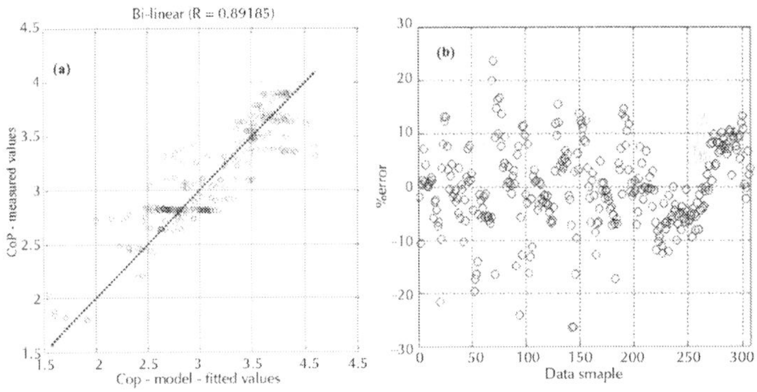

Figure 3: Bi-linear model fitting results. (a): Regression plot; (b): Target-model errors (%).

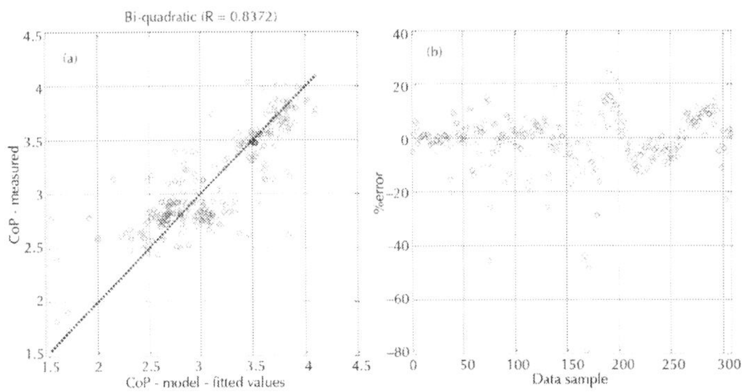

Figure 4: Bi-quadratic model fitting results. (a): Regression plot; (b): Target-model errors (%).

Table 2: Fitted coefficients for the heat pump model

Array Size (m)	A	B	C	D
100	2.852525	2.868282	−0.017015	−0.037951
200	7.936727	−1.206375	−0.154520	0.067780
300	3.315421	2.816345	−0.021494	−0.042534
365	3.188746	3.414232	−0.017961	−0.062144

GROUND ARRAY MODELLING

The closed loop vertical array was modelled as a conventional heat exchanger problem using Kays and London's classical "number of transfer units" (*NTU*) method [18]. This involved calculating the number of heat exchange transfer units from which the array effectiveness could be determined (Equations (4) and (5)):

$$NTU_{array} = \frac{L_{array}}{R_{bhx} m_{f,array} c_{pf,array}} \tag{4}$$

$$E_{array} = 1 - \exp(-NTU) \tag{5}$$

Note that the specific heat capacity of the array fluid, $c_{pf,array}$, is calculated using the properties of ethylene glycol solution at the user-defined ethylene glycol concentration. In the present work, the concentration was set at 20% (by volume). The heat exchange (*i.e.*, array) effectiveness is defined as the heat transfer achieved divided by the maximum theoretically possible heat transfer. Thus, the array outlet temperature can be calculated at each time row in the simulation from Equation (6):

$$T_{f,array,out} = T_{array,soil} + \Delta T_{f,max} \times (1 - 1/E_{array}) \tag{6}$$

(where $T_{array,soil}$ is the current time row average of the soil temperature along the paths of the array heat exchangers).

For the soil domain, a three-dimensional (in space) dynamic numerical solution of the Energy Equation was used through a uniformly discretized grid cast throughout a defined soil domain. Full details of this model can be found in [19]. The house time-series energy demands described in Section 2.2 are read into this model and used to calculate the energy balance on the new heat pump model described in Section 3.3 and the array model described above. This leads to a heat pump source flux at each time row which is then uniformly imposed on the defined array path in the soil domain model. In effect, this amounts to imposing a variable line source of heat on the numerical soil domain model.

The soil domain model used a 50 m × 50 m × 150 m (deep) domain size with a uniform 1 m grid mesh size. The grid mesh size was found to be acceptable for long time horizon simulations (such as in the present work) but a smaller size is needed for short-time simulations (a further discussion about this can be found in [19]). The domain size was found to give good results for up to 5 years of simulation duration. A comparison with a 100 m × 100 m × 150 m domain size was made and differences in results in the key variable of heat pump electricity usage were found to be negligible (the larger domain requiring almost five-times the computation effort as the smaller domain). A uniform undisturbed earth temperature distribution (see Section 5) was imposed throughout the domain as an initial condition and the boundaries of the domain were held at these values. A uniform time step of 1 h was used. The computation time on a quad-core workstation was found to be 4.7 h of simulation time per 1 s of computer elapsed time. The entire ground, heat pump and array model was implemented as a bespoke Matlab function.

RESULTS AND DISCUSSION

Reviews of soil properties based on thermal response tests carried out in the UK have received attention recently. Underwood [19] reported on 13 such tests from a variety of sites giving lower quartile, median and upper quartile soil thermal conductivities of 2.40 $W{\cdot}m^{-1}{\cdot}K^{-1}$, 2.47 $W{\cdot}m^{-1}{\cdot}K^{-1}$ and 3.08 $W{\cdot}m^{-1}{\cdot}K^{-1}$, respectively and lower quartile, median and upper quartile borehole resistances of 0.152 $m{\cdot}K{\cdot}W^{-1}$, 0.162 $m{\cdot}K{\cdot}W^{-1}$ and 0.216 $m{\cdot}K{\cdot}W^{-1}$, respectively. Banks *et al.* [20]

reported on a much larger sample of 61 UK sites. Correspondingly, conductivities were found to be 1.86 $W{\cdot}m^{-1}{\cdot}K^{-1}$, 2.25 $W{\cdot}m^{-1}{\cdot}K^{-1}$ and 3.00 $W{\cdot}m^{-1}{\cdot}K^{-1}$, respectively, and borehole thermal resistances of 0.09 $m{\cdot}K{\cdot}W^{-1}$, 0.11 $m{\cdot}K{\cdot}W^{-1}$ and 0.14 $m{\cdot}K{\cdot}W^{-1}$, respectively [20]. There are some similarities in the conductivity results between the two sources though there are differences between borehole resistance values most likely due to different methods being used to arrive at the results from basic measurements (extraction using a bespoke optimisation algorithm in [19] and extraction by conventional line source theory in [20]). Since [20] represents the larger sample of data, these results will be used in the evaluative modelling that follows. In addition, the undisturbed ground temperatures at the lower quartile, median and upper quartile of 11.7 °C, 12.3 °C and 13.2 °C [20] are used.

Array Design

The median soil property and borehole resistance data reported in [20] were used for initial array design:

Soil thermal conductivity: 2.25 $W \cdot m^{-1} \cdot K^{-1}$

Borehole thermal resistance: 0.11 $m \cdot K \cdot W^{-1}$

Undisturbed soil temperature: 12.3 °C

In addition, the volume heat capacity for the soil was assumed to be 2.4 $MJ{\cdot}m^{-3}{\cdot}K^{-1}$ which is appropriate for a wide range of soil and rock types. Reference is made to the house design heat losses detailed in Appendix A, Table A2; a notional design coefficient of performance of 2.82 as suggested by the Energy Saving Trust's most recent field trials [3]; and the array design look-up tables contained in the Microgeneration Certification Scheme (MCS) [2]. Using this information, recommended array designs of 83 m for the mid-terrace house and 138 m for the detached house were arrived at. These array sizes would therefore be used in normal practice.

For design evaluation, simulations using the foregoing data together with the time-series energy demands detailed in Section 2.2 were carried out for both house types based on a range of array sizes starting with the MCS values of 83 m and 138 m for the respective house types. Results in the form of alternative heat pump seasonal performance factors (*SPF*) are summarized in Table 3. Note that seasonal performance factors used in the present work are what have become referred to in some of the literature as "SPF_{H2}" [3,8]. They are applicable to a monovalent heat pump (including source fluid pump) electricity use but exclude the heating system (sink) pump. For all simulations, the heat pump heating water outlet temperature set point was fixed at 42 °C which is within the range of the test results described in Section 3.2 upon which the heat pump model is based. It was assumed that all space heating and domestic hot water loads are delivered at this temperature from the heat pump buffer store of capacity 50 L. No allowance has been made in the present work for additional direct electric heating due to periodic hot water pasteurization (if used).

Table 3: Design array sizes and corresponding heat pump *SPF*s

Array Size (m)	Mid-Terrace House		Detached House	
	SPF	**Improvement**	***SPF***	**Improvement**
83 (MCS, [2])	3.05	-	-	-
138 (MCS, [2])	-	-	3.02	-
200	3.32	8.9%	3.14	4.0%
300	3.38	1.8%	3.33	6.1%
400	3.35	-	3.31	-

Comments: The performances based on array sizes recommended by the MCS are broadly consistent with the median performance of the actual installations monitored by the Energy Saving Trust [3] after improvements had been made (*i.e.*, 3.1). For both house types, there are improvements of up to around 10% in the heat pump *SPF* by increasing the array size significantly. For the mid-terrace house, increasing the array size by a little over 100 m beyond the value recommended by the MCS gives a performance improvement of 8.9% whereas a further increase by 100 m results in a much lower improvement of 1.8% and there is no improvement at higher array sizes. Therefore, a design array size of 200 m would seem appropriate for this case. For the detached house, there is a performance improvement of 4% when increasing from the MCS recommended array size to 200 m and the improvement increases to 6.1% when increasing from 200–300 m after which there is no further improvement with increasing array size. Thus, a 300 m array would seem to be the appropriate choice here.

In the subsequent analyses, the array size for the mid-terrace house will be set at 200 m and, for the detached house, 300 m.

Sensitivity to Time Horizon

To test the designs over an extended operating period, the array sizes and other design conditions set out in Section 5.1 were applied to five-year simulations for each house. It was assumed that the annual demand patterns (Figure 1) remained the same in each year. Results of the daily minimum array inlet temperatures over the five-year duration are provided in Figure 5 and Figure 6 for both houses. Results of the daily mean heat pump coefficient of performance over the same duration are given in Figure 7 and Figure 8 also for both houses.

Comments: As is to be expected, there is a decline in minimum array fluid temperature over the extended operating time horizon though the rate of decline reduces over time particularly after the first year. What is important with these array designs is that the array fluid temperature for both houses barely falls below 4 °C which suggests that fresh water (rather than ethylene glycol solution) might safely be used with corresponding performance, cost and environmental advantages. The trends in heat pump coefficient of performance (Figure 7 and Figure 8) show a small decline over the five-year period as the ground temperature reduces and a more pronounced decline in the middle period of each year. This is because light loads are being met at these times (*i.e.*, domestic hot water only) with the consequence that the heat pump is operating highly intermittently. (A key assumption in the modelling is that the thermostatically-controlled heat pump operates at the same set point temperature (42 °C) at all times for both heating and hot water delivery and so the heat pump performance is governed by intermittent operation and seasonal variations in the source temperature.) Equally, though the performance is inferior at this time of the year, the heating delivered (and therefore electrical energy consumed by the heat pump) will be lower than in winter and this mitigates the inferior heat pump performance to some extent. The decline in heat pump performance over the five-year horizon is more pronounced with the mid-terrace house than with the detached house and the detached house exhibits a more pronounced dip in summer heat pump performance. The reason for this is that the domestic hot water loads are similar for both houses whereas the detached house has a significantly higher heat load due to space heating. The correspondingly larger array size for the detached house recovers better in summer when demands are light than the mid-terrace house which has the smaller array size. However, a lower pattern of *P* for the detached house due to higher heat load but only marginal increases in summer hot water load results in a lower summer *CoP* than experienced by the mid-terrace house. Again, because relatively low amounts of energy are generated in summer, the inferior heat pump performance is mitigated to some extent.

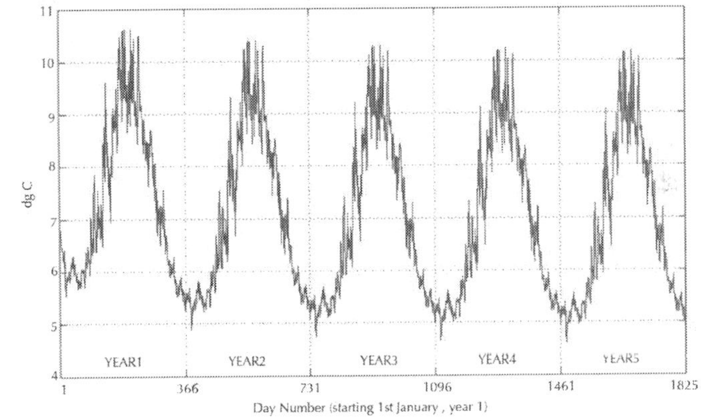

Figure 5: Minimum daily array inlet temperatures (mid-terrace house).

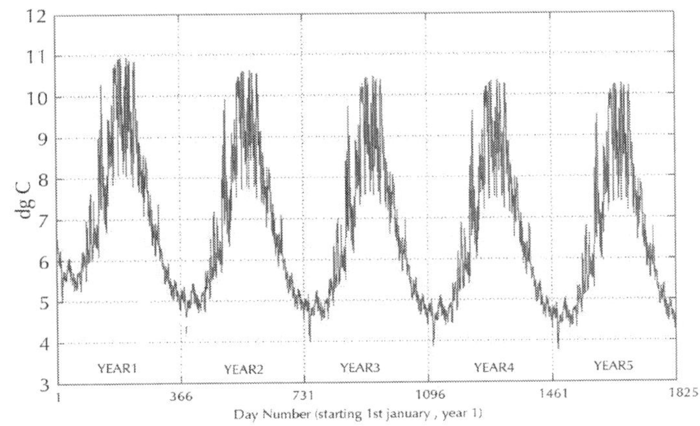

Figure 6: Minimum daily array inlet temperatures (detached house).

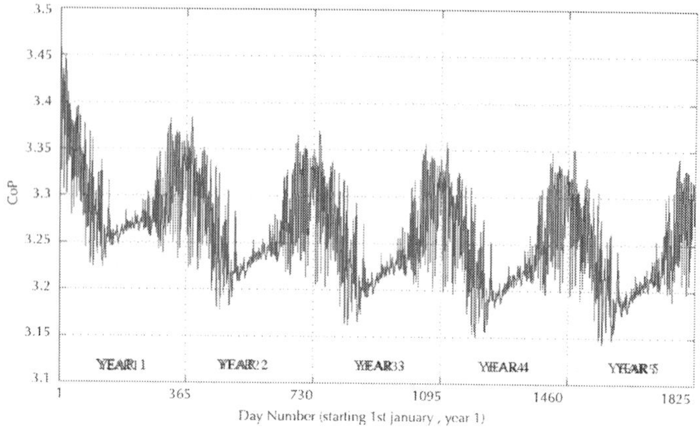

Figure 7: Mean daily heat pump CoP (mid-terrace house).

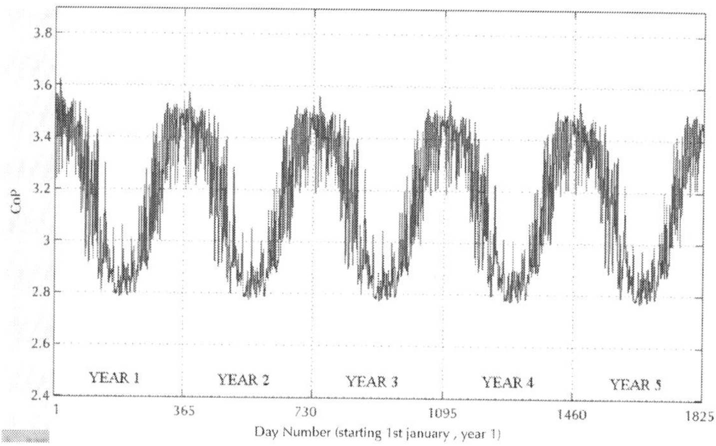

Figure 8: Mean daily heat pump CoP (detached house).

Sensitivity to Soil Conductivity and Array Resistance

To evaluate the sensitivity of the designs to variations in soil conductivity and borehole heat exchanger resistance values, a simulation was conducted using alternative soil conductivities of 1.86 W·m^{-1}·K^{-1} (*i.e.*, the lower quartile value from the Banks *et al.* review [20]) and a further simulation was conducted using this lower conductivity and a higher borehole resistance value of 0.216 m·K·W^{-1}. In the choice of the latter value, the upper quartile resistance from [19] was used instead of from [20] since the range of values in [19] was wider. Results of the original seasonal performance factors compared with the new values arising from these changes in soil and borehole properties are given in Table 4.

Comments: The results show that the performance is not significantly affected by a "typical" range of soil and resistance properties found in UK conditions. Here, the results appear to be more sensitive to borehole resistance than soil conductivity though it should be stressed that this illustration involved a relatively minor reduction in conductivity and a substantial (near doubling) increase in borehole resistance. However, it is becoming clear that the range of soil conductivities across many UK applications is relatively low and this prompts the need to consider carefully whether expensive thermal response testing is needed in every case when ground geology is known with a reasonable degree of confidence. It should also be pointed out that, in most cases, the existence of groundwater flows (not considered in this work) will actually improve performance.

Table 4: Sensitivity to variations in soil conductivity and borehole resistance

Property Choices	***SPF***	
	Mid-Terrace House	Detached House
Median (design) values	3.32	3.33
Reduced k	3.31	3.33
Reduced k & increased R_{bhx}	3.17	3.26

Sensitivity to Heating System Sizing

A final analysis was carried out into the impact of heating system sizing. First, the simulation was re-run with P (in Equation (3)) fixed at a constant value of 1. This will show how much additional energy is required due to the deleterious part-load performance of the heat pump thus revealing the potential for improved control over the heat pump at light loads. Second, the simulation was re-run for both houses with the heating system capacity (Table A2, Appendix A) reduced by 20%, and then increased by 20% and 40%.

Results, compared with the original design results for reference, are given in Table 5. Also included are the averages of the comfort (operative) temperature of the whole house averaged over all periods when the space heating system is active only. Note that the operative temperature as used here is the average of the internal air and mean radiant temperatures.

Table 5: Sensitivity to heating system sizing and comfort

Capacity Options	Mid-Terrace House			Detached House		
	SPF	kWh·m^{-2}	T_{OP} (°C)	*SPF*	kWh·m^{-2}	T_{OP} (°C)
Perfect tracking	3.56	36.1	19.97	3.75	33.0	19.74
Design	3.32	38.6	19.97	3.33	37.1	19.74
20% under-sizing	3.34	35.9	18.68	3.35	34.5	18.46
20% over-sizing	3.31	40.8	20.61	3.31	38.9	20.54
40% over-sizing	3.30	42.4	21.19	3.29	40.3	21.04

(kWh·m^{-2}: annual electricity usage due to heat pump and source fluid pump divided by house gross floor area).

Comments: A 40% heating system over-sizing margin will not translate to a 40% increase in energy because the heating system controls will act to regulate the system to the required comfort conditions. However, oversizing will lead to an increase in energy because an over-sized control system will not be able to track the required control condition perfectly. In particular, the use of simple proportional controls used in domestic "thermostatic" radiator valves will exhibit offset (a sustained difference between set point and actual value). Thus, an over-sized system will lead to an increase in both energy use and comfort temperature. Consequently, the seasonal performance factor is not significantly affected by over-sizing (or under-sizing), however the electricity consumed has increased in all over-sizing cases. If the heat pump was able to operate over all load patterns without loss in performance it would operate with a *SPF* in excess of 3.5 for both houses. This falls to around 3.3 for both houses when its part load behavior is accounted for and the electrical consumption increases by 7% (mid-terrace house) and 12% (detached). If the heating is oversized by up to 40% the energy consumption over a heat pump well matched to the load at all times will increase by almost 18% (mid-terrace house) and 22% (detached house). These should be considered as viable targets for further improvements in domestic heat pump performances in the UK where evidence of equipment over-sizing is plentiful. A particularly notable result in Table 5 is revealed that when the heating system is under-sized by 20% the energy use falls whilst comfort remains very close to acceptable limits. Note here that the area-weighted target comfort temperature for both houses based on 21 °C in the main living space and 18 °C in all other spaces is 18.5 °C—almost precisely met on average by a heating system that is under-sized by 20%. The reason for this is that conventional heating sizing is carried out using steady-state calculations at boundary conditions that, in many winters, will never be realized or if they are, will be of short duration. There is a compelling case for the use of dynamic thermal modelling for the sizing of complex systems such as heat pumps and other microgenerators.

CONCLUSIONS

The aim of this work was to investigate the impact of heat pump capacity and ground array design for domestic ground source heat pumps operating in UK conditions with a view to establishing design criteria that might lead to improvements in operating performance.

The work has been largely based on numerical modelling supported with the introduction of a new empirical heat pump model which takes into account the decline in heat pump performance during periods of light load. Results have been drawn from two exemplar houses which have been configured to be typical of common terraced and detached houses in the UK and use has been made of an existing numerical model of a closed loop vertical ground array.

The results of this work suggest vertical ground loop array sizes for the two typical house types investigated of around 2.5 m of array length per m^2 of house gross floor area. The recommended allowance using the Microgeneration Certification Scheme recommendations would be around 1.1 m/m^2. Furthermore, the array sizes proposed in this work show that it will be possible to operate the array safely using fresh water rather than ethylene glycol solution (or some other form of antifreeze) which is beneficial for performance, cost and the environment.

As data on ground thermal conductivities in the UK start to become more abundant, it is becoming clear that many sites have mean conductivities of around 2–2.5 $W \cdot m^{-1} \cdot K^{-1}$ and modest variations about this figure have little effect on heat pump performance. However, the more uncertain values of borehole heat exchanger resistance do have an influence on performance.

The impact of both deteriorating part-load performance of thermostatically-controlled ground source heat pumps and heating system over-sizing (by up to 40%) has been shown to increase energy use by up to 18%–22% for the two typical house types considered.

Evidence in this work points to a strong potential for a better matching of the capacity of ground source heat pumps to the required building load pattern through the use of dynamic thermal modelling instead of conventional steady-state design methods.

Further work is needed in the following areas:

- Development of verifiably accurate and easy-to-use tools for the design and seasonal performance evaluation of ground source heat pumps suitable for use by practitioners.
- Consideration of the impact of groundwater flow over closed loop arrays in UK conditions.
- Investigations into variable speed drives, electronic expansion devices and improved controls for domestic scale heat pumps.
- Development of alternative dynamically-based methods for system design and capacity-sizing as an alternative to conventional steady-state sizing methods.

Glossary

a1...d1 Regression constants

a2...f2 Regression constants

A...D Regression constants

CoP Coefficient of performance (dimensionless)—instantaneous or short-time average heat output divided by instantaneous or short-time average heat pump plus source fluid pump electrical consumption

$c_{pf,array}$ Specific heat capacity of array fluid ($J \cdot kg^{-1} \cdot K^{-1}$)

E_{array} Array (heat transfer) effectiveness

k Soil mean thermal conductivity ($W \cdot m^{-1} \cdot K^{-1}$)

L_{array} Total array length (m)

$m_{f,array}$ Array fluid mass flow rate ($kg \cdot s^{-1}$)

NTU_{array} Array number of (heat) transfer units

P Part load ratio (current heating demand divided by seasonal maximum heating demand)

R Correlation coefficient

SPF Seasonal performance factor (dimensionless)—seasonal heat pump heating energy divided by seasonal heat pump plus source fluid pump electrical consumption

R_{bhx} Borehole heat exchanger thermal resistance ($m \cdot K \cdot W^{-1}$)

$T_{f,array,out}$ Array fluid outlet temperature (°C)

$T_{array,soil}$ Average soil temperature along the entire array path (°C)

T_{OP} Room space operative temperature (°C)

$\Delta T_{f,max}$ Maximum (design) array inlet/outlet fluid temperature difference (K)

ΔT_{sosi} Heat pump nominal temperature lift (difference between the heating outlet temperature (K or °C) and the source fluid outlet temperature (K or °C)

Symbols used in Table A1 and Table A2

MT mid-terrace

DET detached

GFA gross floor area

LS main living space

B balance of ground floor space

FF first floor

Appendix A. Seasonal Energy Demand Modelling

Table A1: Seasonal house energy modeling—main parameters

Type	Areas (m^2)		Footprint		Orientation	Windows	Glazing (m^2 *)	Roof
	GFA	**LS**	**Width (m)**	**Depth (m)**				
MT	75	12	6.1	6.1	N-S	Clear double	10.4	Pitched
DET	125	20	7.9	7.9	N-S	Clear double	17.2	Pitched

* **Based on the UK Standard Assessment Procedure: 0.1382** × *GFA* − 0.027. An additional allowance of one single door (rear) and one single door (front) of 2 m^2 **(each) is added to the window areas.**

Table A2: Seasonal house energy modeling—thermal properties

U-values ($W \cdot m^2 \cdot K^{-1}$)				**Design Temperatures (°C)**			**Ventilation (h^{-1})**	**Design Loss (kW *)**
Wall	**Roof**	**Gr-floor**	**Window**	**LS**	**B**	**FF**		
0.45	0.35	0.45	2.2	21	18	18	0.5	3.09
0.45	0.35	0.45	2.2	21	18	18	0.5	5.95

* **Design heat loss with an external design temperature of** −3 °C. Calculations include a pre-heat margin of 25% and were carried out in accordance with the methods set out in the CIBSE Guide. The standards are those that would be expected of UK houses either constructed or refurbished at around 2002.

Appendix B. Dynamic Thermal Modelling—Simulink HVAC Blockset

The Simulink HVAC Blockset is a generic Simulink library that can be used to construct detailed dynamic models of buildings including HVAC plant and controls and certain embedded renewable energy systems. Several such libraries exist for modelling energy in buildings such as SIMBAD [21] and CARNOT [22] but the advantage of the HVAC Blockset used here is that all component models are fully dynamic, enabling more accurate control and system response modelling to be carried out. (The HVAC Blockset used here is made freely available by the author for other users.) Note that only certain component model selections were used in the present work as detailed below.

Components used to remodel the two house types:

Building envelope:	Plant and controls:	Utility:
Zone heat balance	Generic emitter [2]	Schedule [4]
Exposed element	Detector [3]	Solar simulator [5]
Internal element	Control valve [3]	
Window	PID controller [3]	
Ventilation [1]		

Infiltration due to wind and stack effect during winter with closed windows. The block was set to open windows by 50% of their opening capacity when internal temperatures reached 26 °C and by a further 50% to fully-open when temperatures reached 28 °C (*i.e.*, in summer when the heating is off).

1. The "generic emitter" was used to represent panel convector-radiator heating.
2. The "detector", "control valve" and "PID controller" blocks were combined to represent "thermostatic" radiators valves attached to each zone radiator. Only the proportional term of the PID controller block was enacted. The detector was set with a long time constant (3 min) to help represent the long time delay in these valves. (Note that the widely-used term "thermostatic" is a misnomer in this context since these control valves actually modulate the hot water flow rate in practice.)
3. Separate "schedule" blocks were used to represent switching of occupant activity and plant activity.
4. The "solar simulator" was used to generate in-plane irradiances on each exposed window and opaque surface.

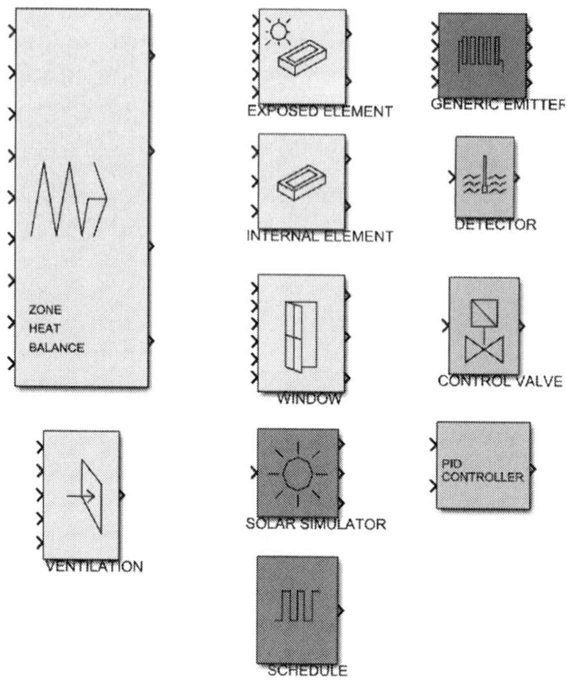

Figure B1: Simulink HVAC Blockset.

The highest level of the Simulink model created for the mid-terrace house is shown as an example in Figure B2, showing all parent blocks and information flows between them. The three main blocks at the centre represent the parent blocks of the living room (top), balance of ground floor spaces (middle) and first floor (bottom). Beneath these parent blocks are all the component blocks (and their information flow connections) for the room elements, zone energy balance, ventilation, heating system and heating controls. The large block to the left in Figure B2 is the parent block which contains all microclimate modelling (*i.e.*, a file-read utility which reads in weather data and solar simulating blocks for all orientations forming the mid-terrace house).

For further details of the individual block descriptions and derivations, see [12,13].

Note that the other blocks shown in the generic library and not specifically referred to above were not used in the present work. Furthermore, the heat pump model in the generic library (which is a simple manufacturer's catalogue-fit type model) was not used in the present work either.

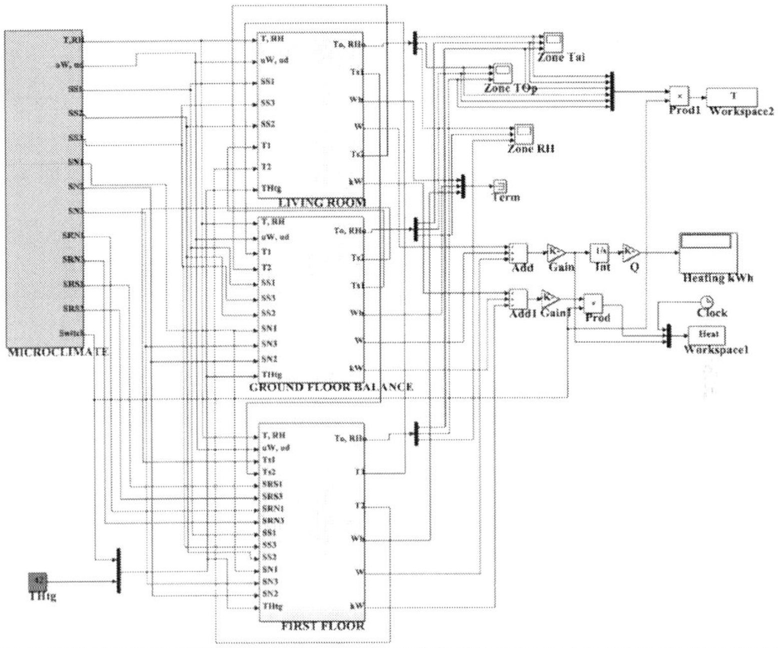

Figure B2: Top-level block diagram model for the mid-terrace house.

Appendix C. Details of the Test Ground Source Heat Pump

Table C1: Details of the test heat pump

System Component	Parameter	Value
Compressor	Type	Scroll
	Refrigerant	R410A
	Electrical supply	1-phase, 220 V
	Displacement	5.34 $m^3 \cdot h^{-1}$

Evaporator and condenser	Type	Brazed plate
	Plate material	Corrugated stainless steel
	Number of plates	13
	Plate width	118.4 mm
	Plate height	440 mm
	Plate spacing	2.24 mm
	Plate thickness	0.4 mm
	Volume, refrigerant side	0.57 L
	Volume, water side	0.66 L
Expansion device	Type	Mechanical, thermostatic

REFERENCES

1. Getting Warmer: A Field Trial of Heat Pumps. Available online: http://www.heatpumps.org.uk/PdfFiles/TheEnergySavingTrust-GettingWarmerAFieldTrialOfHeatPumps.pdf (accessed on 11 December 2013).
2. Microgeneration Installation Standard MIS 3005—Requirements for Contractors Undertaking the Supply, Design, Installation, Set to Work, Commissioning and Handover of Microgeneration Heat Pump Systems, Issue 4.0. Available online: http://www.microgenerationcertification.org/images/MIS%203005%20Issue%204.0%20Heat%20Pump%20Systems%20%202013.12.16%20FINAL3.pdf (accessed on 25 April 2014).
3. The Heat is on. Available online: http://www.energysavingtrust.org.uk/Organisations/Technology/Field-trials-and-monitoring/Fieldmonitoring/Field-trial-reports/Heat-pump-field-trials (accessed on 25 April 2014).
4. Bagdanavicius, A.; Jenkins, N. Power requirements of ground source heat pumps in a residential area. *Appl. Energy* 2013, *101*, 591–600. [Google Scholar] [CrossRef]
5. Blanco, D.L.; Nagano, K.; Morimoto, M. Experimental study on

a monovalent inverter-driven water-to-water heat pump with a desuperheater for low energy houses. *Appl. Therm. Eng.* 2013, *50*, 826–836.

6. Wood, C.J.; Liu, H.; Riffat, S.B. An investigation of the heat pump performance and ground temperature of a pile foundation heat exchanger system for a residential building. *Energy* 2010, *35*, 4932–4940.
7. Boait, P.J.; Fan, D.; Stafford, A. Performance and control of domestic ground-source heat pumps in retrofit installations. *Energy Build.* 2011, *43*, 1968–1976.
8. Gleeson, C.P.; Lowe, R. Meta-analysis of European heat pump field trial efficiencies. *Energy Build.* 2013, *66*, 637–647.
9. Fernández-Seara, J.; Pereiro, A.; Bastos, S.; Dopazo, J.A. Experimental evaluation of a geothermal heat pump for space heating and domestic water simultaneous production. *Renew. Energy* 2012, *48*, 482–488.
10. National Energy Efficiency Data-Framework (NEED), Table 1. Available online: https://www.gov.uk/government/collections/national-energy-efficiency-data-need-framework (accessed on 28 April 2014).
11. BREDEM 2010—A Technical Description of the BRE Domestic Energy Model, Version 1.0. Available online: http://www.bre.co.uk/filelibrary/bredem/BREDEM-2012-specification.pdf (accessed on 28 April 2014).
12. Gouda, M.M.; Danaher, S.; Underwood, C.P. Building thermal model reduction using nonlinear constrained optimization. *Build. Environ.* 2002, *37*, 1255–1265.
13. Gouda, M.M.; Underwood, C.P.; Danaher, S. Modelling the robustness properties of HVAC plant under feedback control. *Build. Serv. Eng. Res. Technol.* 2003, *24*, 271–280.
14. CIBSE TM48. *Use of Climate Change Scenarios for Building Simulation: The CIBSE Future Weather Years*; Chartered Institution of Building Services Engineers: London, UK, 2009.
15. Underwood, C.P.; Yik, F.W.H. *Modelling Methods for Energy in Buildings*; Blackwell: Oxford, UK Chapter 4. , 2004; pp. 135–137.

16. Fisher, D.E.; Rees, S.J.; Padhmanabhan, S.K.; Murugappan, A. Integration and validation of ground source heat pump system models in an integrated building and system simulation environment. *HVAC&R Res.* 2006, *12*, 693–710.
17. Kinab, E.; Marchio, D.; Rivière, P.; Zonghaib, A. Reversible heat pump model for seasonal performance optimization. *Energy Build.* 2010, *42*, 2269–2280.
18. Kays, W.M.; London, A.L. *Compact Heat Exchangers*, 2nd ed.; McGraw-Hill: New York, NY, USA, 1964.
19. Underwood, C.P. Ground-source heat pumps: Observations from United Kingdom ground thermal response tests. *Build. Serv. Eng. Res. Technol.* 2013, *34*, 123–144.
20. Banks, D.; Withers, J.G.; Cashmore, G.; Dimelow, C. An overview of the results of 61 *in situ* thermal response tests in the UK. *Q. J. Eng. Geol. Hydrogeol.* 2013, *46*, 281–291.
21. SIMBAD—Simulation of Buildings and Devices. Available online: http://www.simbad-cstb.fr/ (accessed on 20 June 2014).
22. CARNOT—Conventional and Renewable Energy Siystems Optimisation Blockset. Available online: http://mv.fh-duesseldorf.de/d_pers/Adam_Mario/a_lehre/gm_allg_down/Carnot-Hilfe.pdf (accessed on 20 June 2014).

Chapter 4

A Rolling Constraint Reproduces Ground Reaction Forces and Moments in Dynamic Simulations of Walking, Running, and Crouch Gait

Samuel R. Hamner,[a] Ajay Seth,[b] Katherine M. Steele,[a] and Scott L. Delp[a,b,*]

[a]Department of Mechanical Engineering, Stanford University, USA
[b]Department of Bioengineering, Stanford University, USA

ABSTRACT

Recent advances in computational technology have dramatically increased the use of muscle-driven simulation to study accelerations produced by muscles during gait. Accelerations computed from muscle-driven simulations are sensitive to the model used to represent contact between the foot and ground. A foot-ground contact model must be able to calculate ground reaction forces and moments that are consistent with experimentally measured ground reaction forces and moments. We show here that a rolling constraint can model foot-ground contact and reproduce measured ground reaction forces and moments in an induced acceleration analysis of muscle-driven simulations of walking, running, and crouch gait. We also illustrate that a point constraint and a weld constraint used to model foot-ground contact in previous studies produce inaccurate reaction moments and lead to contradictory interpretations of muscle function. To enable others to use and test these different constraint types (i.e., rolling, point, and weld constraints) we have included them as part of an induced acceleration analysis in OpenSim, a freely-available biomechanics simulation package.

INTRODUCTION

Muscle-driven simulations of human gait have provided insights into the actions of muscles during walking (e.g., Anderson and Pandy, 2003; Liu et al., 2008), running (e.g., Hamner et al., 2010; Sasaki and Neptune, 2006), and pathological gait (e.g., Peterson et al., 2010; Steele et al., 2010). These studies employ methods, like induced acceleration analysis, which decompose ground reaction forces and moments using foot-ground contact models to examine how muscle forces contribute to accelerations of joints and the body mass center. Results from these studies depend upon the ground contact model (Dorn et al., 2012a; Hamner et al., 2010). It is therefore essential to assess the ability of different contact models to produce accurate ground reactions. Previous studies have compared simulated reaction forces to experimentally measured ground reaction forces (Anderson and Pandy, 2001; Dorn et al., 2012b; Seth and Pandy, 2007). It is also necessary to compare simulated reaction *moments* to measured ground reaction *moments*.

The purpose of this study was to examine the accuracy with which a rolling constraint represents contact between the foot and ground during induced acceleration analyses of walking, running, and crouch gait. Induced acceleration analysis uses a model of foot-ground contact to determine how muscles, gravity, and velocity-related forces contribute to the ground reaction force. To test the accuracy of a rolling constraint we compared ground reaction forces and moments computed in an induced acceleration analysis to experimentally measured ground reaction forces and moments. We also evaluated other constraint-based contact models in an induced acceleration analysis of running to illustrate that these different models produce inaccurate reaction moments and lead to contradictory interpretations of muscle function.

METHODS

We used simulations of three different gait patterns: walking (Liu et al., 2008), running (Hamner et al., 2010), and crouch gait (Steele et al., 2012) (Fig. 1). Marker trajectories and ground reaction forces and moments were measured while subjects either walked over ground or ran on a treadmill instrumented with force plates. Muscle-driven simulations were generated from these experimental data using OpenSim (Delp et al., 2007). A musculoskeletal model consisting of 92 muscles of the lower extremities and torso was scaled to match each subject's anthropometry using experimentally measured markers placed on anatomical landmarks and calculated joint centers. A corresponding virtual marker set was placed on the model based on these anatomical landmarks. Joint angles were calculated using an inverse kinematics algorithm that minimized the difference between experimental and virtual markers at each time frame (Delp et al., 2007). The computed muscle control algorithm (Thelen et al., 2003) determined muscle excitation patterns required to track measured motion. Details of experimental and simulation methods are included in the primary publications (Hamner et al., 2010; Liu et al., 2008; Steele et al., 2012).

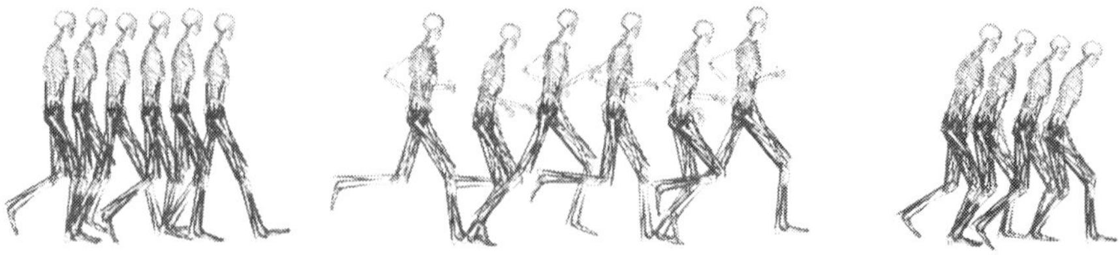

Figure 1: Simulations of a walking gait cycle, a running gait cycle, and a crouch gait cycle. The simulations shown are of a representative subject from each study. Each set of simulations (i.e., walking, running, crouch) consisted of data from three subjects; all were analyzed using the rolling constraint to model ground contact during an induced acceleration analysis. Each simulation used a scaled musculoskeletal model with the lower extremities and torso driven by 92 musculotendon actuators.

An induced acceleration analysis was used to compute reaction forces and moments of a constraint between the foot and ground due to the forces acting on the musculoskeletal model: muscle forces, gravity, and forces due to velocity effects (i.e., Coriolis and centripetal forces). The resultant constraint reaction force and moment were calculated by summing the constraint reactions due to each force acting on the musculoskeletal model. This calculated sum was then compared to the experimentally measured ground reaction force and moment to assess accuracy.

We calculated accelerations due to forces acting on the musculoskeletal model using equations of motion for a constrained, rigid body system with muscles (Sherman et al., 2011):

$$[M]\ddot{q}+[C]^{T}\lambda=[R]F_{m}+F_{g}+F_{v} \quad (1)$$

$$[C]\ddot{q}=b \quad (2)$$

where M is the mass matrix, q is a vector of generalized coordinates (e.g., joint angles), C is the constraint matrix, λ reprsents generalized constraint reaction forces, F_m represents muscle forces, F_g represents gravitational forces, F_v represents forces due to velocity effects, R

is the matrix of muscle moment arms, and$b$ is a vector containing position and velocity terms (i.e., q and $\dot{q}$) of the constraint equations (i.e., Eqs. (3)–(6)) expressed in terms of generalized accelerations, $\ddot{q}$. The constraint matrix C maps generalized constraint reaction forces λ to generalized forces. Generalized constraint reaction forces λ and generalized accelerations$\ddot{q}$ are solved simultaneously.

To model contact between the foot and ground, we implemented a rolling constraint (Fig. 2; ROLL) that generates a fore-aft, vertical, and mediolateral reaction force and a vertical reaction moment. With a constraint-based contact model, constraint equations are included in the equations of motion and reactions are computed at each time step. Contact models utilizing spring-dampers (e.g., Anderson and Pandy, 2001;Neptune et al., 2000) require equations of motion to be integrated forward in time to calculate contact forces. Constraint-based models eliminate the computational cost of forward integrations and sensitivity of results due to differing integration time windows.

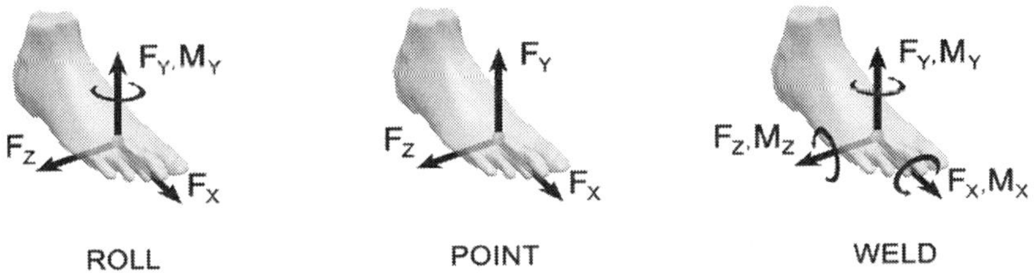

Figure 2: The rolling constraint (ROLL) generates a fore-aft (F_X), vertical (F_Y), and mediolateral (F_Z) reaction force and the vertical reaction moment (M_Y) about the center of pressure. The POINT constraint can only generate a fore-aft, vertical, and mediolateral reaction force. The WELD constraint can generate a fore-aft, vertical, and mediolateral reaction force and a frontal (M_X), vertical (M_Y), and sagittal (M_Z) reaction moment about the center of pressure.

The rolling constraint combines four individual constraints: a unilateral non-penetrating constraint (Eq. (3); i.e., foot cannot penetrate ground but can be lifted) and a pure rolling constraint (Kane, 1961) (Eqs. (4)–(6)), which includes two no-slip constraints (i.e., limits fore-aft and mediolateral foot translations) and a no-twist constraint about the axis normal to ground (i.e., limits vertical rotations). Each constraint is applied at the measured center of pressure.

Constraint name	Equation	Constraint condition(s)
Vertical, unilateral, non-penetrating	$\rho_Y(q) \geq 0$	$F_Y > F_{threshold}$ (3)
Fore-aft no slip	$v_X(q, \dot{q}) = 0$	Non-penetrating condition and $F_{Y\mu friction} > \sqrt{F_X^2 + F_Z^2}$ (4)
Mediolateral no slip	$v_Z(q, \dot{q}) = 0$	non-penetrating condition and $F_{Y\mu friction} > \sqrt{F_X^2 + F_Z^2}$ (5)
Vertical no-twist	$\omega_Y(q, \dot{q}) = 0$	no slip condition and $r_{contact}\mu_{friction}F_Y > M_Y$ (6)

In Eqs. (3)–(6), ρ_Y is the vertical position of the foot, v_X and v_Z are the fore-aft and mediolateral foot velocities, respectively, and ω_Y is the foot's vertical angular velocity, all with respect to ground. $\hat{F}_Y$ is the measured vertical ground reaction force, F_x and F_z are the simulated fore-aft and mediolateral constraint reaction forces, respectively, and M_Y is the simulated vertical constraint reaction moment. The constraint equations are differentiated to provide constraints on the system accelerations, $\ddot{q}$ (Eq. (2)).

Specified parameters were used to determine when each constraint was active (i.e., turn on/off each constraint) based on constraint conditions in Eqs. (3)–(6). If conditions were met, each constraint was applied to the foot at the measured center of pressure, allowing direct comparison of simulated constraint reactions and measured ground reactions. In the constraint conditions, $F_{threshold}$ is a threshold (5 N) for the vertical reaction force used to determine when the non-penetrating constraint is active (Eq. (3)), $\mu_{friction}$ is a friction coefficient (0.65) used to determine when the no-slip constraints are active (Eqs. (4)–(5)), and $r_{contact}$ is a contact radius (0.01 m) representing the size of the contact area between foot and ground and is used to determine a threshold for the reaction moment (Eq. (6)). Parameter values were determined by varying each parameter within a range of physically realistic values (i.e., 0 N < $F_{threshold}$ < 50 N; 0.01 < $\mu_{friction}$ < 1 and 0.001 m < $r_{contact}$

< 0.1 m) and selecting values that provided appropriate timing for heel strike and toe-off. Parameters only affected when the constraints were active and varying parameters had no effect on the magnitude of reaction forces and moments calculated with the constraints. The accuracy of the rolling constraint was assessed by calculating root-mean square (RMS) difference between each component of measured ground reactions and simulated constraint reactions, averaged from the three subjects in each study.

To examine how different constraint-based contact models affect interpretation of muscle function, we conducted a case study using running simulations in which we quantified how different constraints affect muscle contributions to mass center accelerations calculated by induced acceleration analysis. We examined running as it produces larger, more rapidly changing ground reaction forces than walking. We compared the rolling constraint with two constraint-based contact models used in previous studies: a point constraint (Fig. 2; POINT) and a weld constraint (Fig. 2; WELD). The point constraint does not allow the foot to translate in any direction (i.e., fore-aft, mediolateral, or vertical), while it allows the foot to rotate about all three axes (e.g., Kepple et al., 1997). Thus, a point constraint applied at the center of pressure generates reaction force in all directions (i.e., F_x, F_y, and F_z), but cannot generate any reaction moment (i.e., M_x, M_y, and M_z). The weld constraint does not allow the foot to translate or rotate (e.g., Anderson and Pandy, 2001) and can thus generate reaction forces and moments in all directions. By comparison, the rolling constraint limits foot translation in all directions while only limiting rotation about the vertical axis, thus generating reaction forces in all directions but only a vertical reaction moment. Each constraint (ROLL, POINT, and WELD) was applied in the same induced acceleration analysis framework, allowing for direct comparison between simulated constraint reactions and measured ground reactions. The point and weld constraints were turned on when the vertical ground reaction force exceeds a specified threshold (i.e., $\hat{\mathbf{F}}_Y > F_{threshold}$). Accuracy of each constraint type was assessed by calculating RMS difference between each component of average simulated constraint reactions and average measured ground reactions. We also calculated average contributions of soleus and vasti to fore-aft and upward mass center accelerations during stance, as these represent important muscle groups for braking and propulsion (Hamner et al., 2010; Liu et al., 2008; Neptune et al., 2008).

RESULTS

The rolling constraint produced reaction forces and moments similar to ground reaction forces and moments measured about the center of pressure for walking, running and crouch gait (Fig. 3). In each case, muscles of the lower extremities and torso, gravity, and forces due to velocity effects accounted for more than 95% of the ground reaction force and moment. The rolling constraint, by design, only produced a vertical reaction moment when applied at the center of pressure. Analysis of running produced the largest difference between simulated constraint reactions and measured ground reactions, with a maximum RMS difference of 0.08 body weights in vertical reaction force and 0.04 Nm/kg in vertical reaction moment (Table 1).

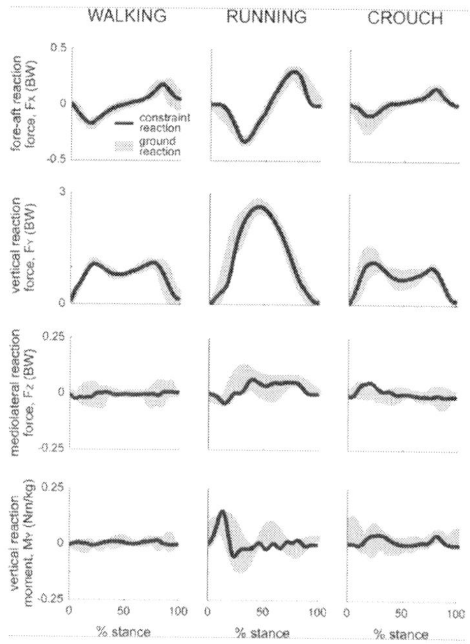

Figure 3: Comparison of the simulated constraint reaction forces and moments (black) with the experimentally measured ground reaction forces and moments (gray) for walking, running, and crouch gait using the rolling constraint (ROLL). The shaded region represents the mean ± one standard deviation of the measured ground reactions for three subjects included in each study.

Table 1: RMS difference between each component of the simulated constraint reaction and experimentally measured ground reaction force and moment, as calculated about the center of pressure, for three different types of gait (walking, running, and crouch) using the rolling (ROLL) constraint. The ground reaction force is normalized by body weights (BW), and the ground reaction moment is normalized by body mass (Nm/kg). Note that the RMS difference of M_X and M_Z for the rolling constraint during walking, running, and crouch is essentially zero ($\lll$0.01)

Gait type	F_X (BW)	F_Y (BW)	F_Z (BW)	M_X (Nm/kg)	M_Y (Nm/kg)	M_Z (Nm/kg)
WALKING	0.01	0.03	0.01	$\lll$0.01	0.01	$\lll$0.01
RUNNING	0.04	0.08	0.02	$\lll$0.01	0.04	$\lll$0.01
CROUCH	0.03	0.03	0.02	$\lll$0.01	0.02	$\lll$0.01

The point constraint reproduced measured ground reaction forces with a maximum RMS difference of 0.09 body weights in vertical reaction force (Table 2), but did not generate any reaction moment because it does not constrain foot rotation (Fig. 4, POINT). In contrast, the weld constraint generated reaction moments about all three axes because it constrains foot rotation in all directions (Fig. 4, WELD). Before toe-off, as the foot rotates, the weld constraint generated large, inaccurate forces and moments, with a maximum RMS difference of 1.0 body weight in mediolateral force and 8.0 Nm/kg in frontal reaction moment (Table 2).

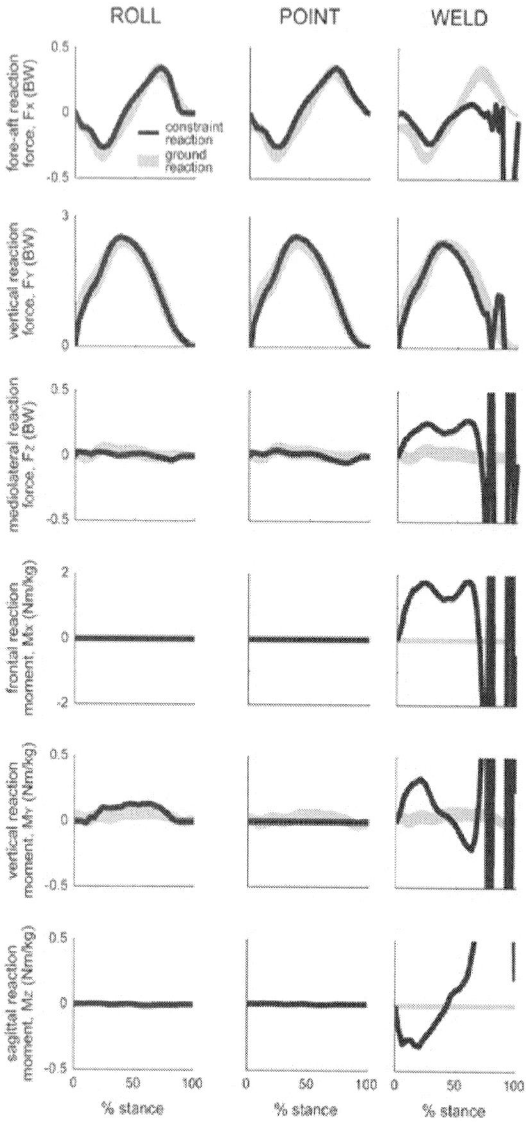

Figure 4: Comparison of the simulated constraint reaction forces and moments (black) with the experimentally measured ground reaction forces and moments (gray) during simulations of a single subject running using three different constraint types (ROLL, POINT, WELD; see Fig. 2). The shaded region represents the mean ± one standard deviation of the measured ground reactions over 14 strides.

Table 2: RMS difference between each component of the simulated constraint reaction and experimentally measured ground reaction force and moment, as calculated about the center of pressure in simulations of a single subject running over 14 consecutive gait cycles using three different constraint types (ROLL, POINT, WELD; see Fig. 2). The ground reaction force is normalized by body weights (BW) and the ground reaction moment is normalized by body mass (Nm/kg). Note that the RMS difference of M_X and M_Z for the ROLL and POINT constraints is essentially zero ($\ll$0.01)

Constraint type	F_X (BW)	F_Y (BW)	F_Z (BW)	M_X (Nm/kg)	M_Y (Nm/kg)	M_Z (Nm/kg)
ROLL	0.03	0.08	0.01	$\ll$0.01	0.02	$\ll$0.01
POINT	0.03	0.09	0.01	$\ll$0.01	0.03	$\ll$0.01
WELD	0.28	0.28	1.00	7.99	3.23	1.51

Using different constraints led to different interpretations of muscle function during running. With the rolling and point constraints, vasti was the largest contributor to backward and upward mass center accelerations during early stance, while soleus was the largest contributor to forward and upward mass center acceleration during late stance. In contrast, roles of vasti and soleus were reversed with the weld (Fig. 5), similar to results observed by Dorn et al. (2012a). Average soleus contribution to upward acceleration also diminished by about 3 m/s^2 compared to the rolling or point constraints, while average vasti contribution increased comparably (Fig. 5). The rolling and point constraints produced similar mass center accelerations.

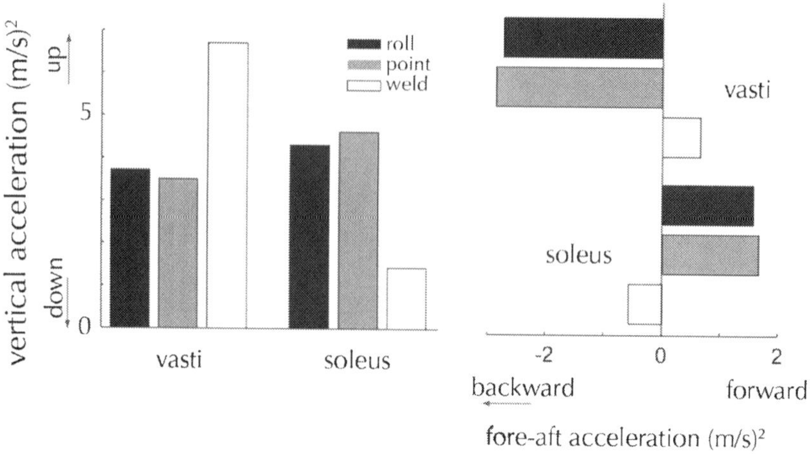

Figure 5: Comparison of the average contributions of vasti and soleus to upward mass center acceleration (i.e., support) and fore-aft mass center accelerations (i.e., braking and propulsion) during the stance phase of running for three different constraint types (ROLL, POINT, WELD; see Fig. 2) in simulations of a single subject running over 14 consecutive gait cycles.

DISCUSSION

The rolling constraint accurately reproduced ground reaction forces and moments during induced acceleration analysis of walking (Liu et al., 2008), running (Hamner et al., 2010), and crouch gait (Steele et al., 2012). In the case of running, the rolling constraint reproduced experimentally measured ground reaction forces and moments more accurately than the point and weld constraints. The rolling constraint produced a moment about the vertical axis because only vertical foot rotation is constrained with respect to ground. Other contact models that limit foot rotations about the frontal and/or sagittal axes (e.g., weld or multiple point constraints) or allow foot rotation about the vertical axis (e.g., point constraint) cannot reproduce the measured ground reaction moment. The vertical ground reaction moment has been shown to significantly affect arm swing during walking and running (Collins et al., 2009; Hinrichs, 1987; Li et al., 2001; Park, 2008), and can affect muscle contributions to mass center and lower extremity joint accelerations through dynamic coupling. Additionally, accurately reproducing the vertical ground reaction moment may be important for understanding pathological gaits, where ground reaction moments are less consistent between subjects. We observed that the weld constraint produces reaction moments much greater than experimentally measured moments and yields vasti and soleus contributions to braking and propulsion in the opposite direction of contributions observed with the rolling or point constraints in analysis of running. Furthermore, the rolling constraint does not require non-physical transition functions necessitated when multiple point constraints are used (Anderson and Pandy, 2003; Dorn et al., 2012a).

As with any model, the rolling constraint has limitations to its accuracy. The constraint utilizes the measured ground reaction force and center of pressure, which will include any errors associated with these experimental measurements. For example, inaccuracy in determining the center of pressure location can occur at the beginning and end of stance as the ground reaction force approaches zero. Additionally, by utilizing the measured center of pressure, the rolling constraint is limited to analyses of measured data and cannot be used in predictive simulations (e.g., Wang et al., 2012). The rolling constraint also represents a rigid connection between foot and ground, which fails to capture deformation of soft tissues during the impact of landing. Therefore, induced acceleration analysis results should be carefully examined during periods of rapid loading and unloading and further testing of rigid constraints should be performed when analyzing movements with impacts, like jumping or cutting. Our models also represent the foot as a single rigid segment, when in reality the foot is a compliant, multiarticular structure. This single segment model is useful for analysis of full-body motions and mass center accelerations, but detailed studies of the foot may require a more complex model.

To have confidence in evaluations of muscle actions derived from muscle-driven simulations, ground contact models should generate forces and moments similar to those measured experimentally. To allow other researchers to assess accuracy of different constraint types we have included an induced acceleration analysis tool with various constraints (e.g., rolling, weld, and point) in OpenSim (Delp et al., 2007; Seth et al., 2011), open source musculoskeletal simulation software. Additionally, to allow others to reproduce and build upon the results of this study, models and data analyzed are freely-available to download from www.simtk.org.

ACKNOWLEDGMENTS

We thank Chand John, Jill Higginson, May Liu, and Mike Schwartz for collecting data, creating simulations, and making the results freely available. Samuel Hamner was funded by fellow-ships from Stanford and NSF. This work was also supported by NIH grants U54 GM072970, R24 HD065690, and R01 HD033929.

REFERENCES

1. Anderson FC, Pandy MG. Dynamic optimization of human walking. Journal of Biomechanical Engineering. 2001;123:381–390.
2. Anderson FC, Pandy MG. Individual muscle contributions to support in normal walking. Gait & Posture. 2003;17:159–169.
3. Collins SH, Adamczyk PG, Kuo AD. Dynamic arm swinging in human walking. Proceedings of the Royal Society B: Biological Sciences. 2009;276:3679–3688.
4. Delp SL, Anderson FC, Arnold AS, Loan P, Habib A, John CT, Guendelman E, Thelen DG. OpenSim: open-source software to create and analyze dynamic simulations of movement. IEEE Transactions on Bio-Medical Engineering. 2007;54:1940–1950.
5. Dorn TW, Lin YC, Pandy MG. Estimates of muscle function in human gait depend on how foot-ground contact is modelled. Computer Methods in Biomechanics and Biomedical Engineering.2012a;15:657–668.
6. Dorn TW, Schache AG, Pandy MG. Muscular strategy shift in human running: dependence of running speed on hip and ankle muscle performance. Journal of Experimental Biology.2012b;215:1944–1956.
7. Hamner SR, Seth A, Delp SL. Muscle contributions to propulsion and support during running.Journal of Biomechanics. 2010;43:2709–2716.
8. Hinrichs RN. Upper extremity function in running II: angular momentum considerations.International Journal of Sport Biomechanics. 1987;3:242–263.

9. Kane TR. Analytical Elements of Mechanics: Dynamics. Vol. 2. New York: Academic Press; 1961.
10. Kepple TM, Siegel KL, Stanhope SJ. Relative contributions of the lower extremity joint moments to forward progression and support during gait. Gait & Posture. 1997;6:1–8.
11. Li Y, Wang W, Crompton RH, Gunther MM. Free vertical moments and transverse forces in human walking and their role in relation to arm-swing. The Journal of Experimental Biology. 2001;204:47–58.
12. Liu MQ, Anderson FC, Schwartz MH, Delp SL. Muscle contributions to support and progression over a range of walking speeds. Journal of Biomechanics. 2008;41:3243–3252.
13. Neptune RR, Sasaki K, Kautz SA. The effect of walking speed on muscle function and mechanical energetics. Gait & posture. 2008;28:135–143.
14. Neptune RR, Wright IC, van den Bogert AJ. A method for numerical simulation of single limb ground contact events: application to heel–toe running. Computer Methods in Biomechanics and Biomedical Engineering. 2000;3:321–334.
15. Park J. Synthesis of natural arm swing motion in human bipedal walking. Journal of Biomechanics.2008;41:1417–1426.
16. Peterson CL, Hall AL, Kautz SA, Neptune RR. Pre-swing deficits in forward propulsion, swing initiation and power generation by individual muscles during hemiparetic walking. Journal of Biomechanics. 2010;43:2348–2355.
17. Sasaki K, Neptune RR. Differences in muscle function during walking and running at the same speed. Journal of Biomechanics. 2006;39:2005–2013.
18. Seth A, Pandy MG. A neuromusculoskeletal tracking method for estimating individual muscle forces in human movement. Journal of Biomechanics. 2007;40:356–366.
19. Seth A, Sherman M, Reinbolt JA, Delp SL. OpenSim: a musculoskeletal modeling and simulation framework for in silico investigations and exchange. Procedia IUTAM. 2011;2:212–232.
20. Sherman MA, Seth A, Delp SL. Simbody: multibody dynamics for biomedical research. Procedia IUTAM. 2011;2:241–261.

21. Steele KM, Seth A, Hicks JL, Schwartz MH, Delp SL. Muscle contributions to vertical and fore-aft accelerations are altered in subjects with crouch gait. Gait & Posture.http://dx.doi.org/10.1016/j.gaitpost.2012.10.019, in press.
22. Steele KM, Seth A, Hicks JL, Schwartz MS, Delp SL. Muscle contributions to support and progression during single-limb stance in crouch gait. Journal of Biomechanics. 2010;43:2099–2105.
23. Thelen DG, Anderson FC, Delp SL. Generating dynamic simulations of movement using computed muscle control. Journal of Biomechanics. 2003;36:321–328.
24. Wang JM, Hamner SR, Delp SL, Koltun V. Optimizing locomotion controllers using biologically-based actuators and objectives. ACM Transactions on Graphics Proceedings of SIGGRAPH.2012; 2012:31.

Chapter 5

The Effect Analysis of Strain Rate on Power Transmission Tower-Line System under Seismic Excitation

Li Tian, [1,*] Wenming Wang, [2] and Hui Qian [3]

[1]School of Civil and Hydraulic Engineering, Shandong University, Jinan 250061, China

[2]Shandong Electric Power Engineering Consulting Institute Co., Ltd., Jinan 250013, China

[3]School of Civil Engineering, Zhengzhou University, Zhengzhou 450001, China

ABSTRACT

The effect analysis of strain rate on power transmission tower-line system under seismic excitation is studied in this paper. A three-dimensional finite element model of a transmission tower-line system is created based on a real project. Using theoretical analysis and numerical simulation, incremental dynamic analysis of the power transmission tower-line system is conducted to investigate the effect of strain rate on the nonlinear responses of the transmission tower and line. The results show that the effect of strain rate on the transmission tower generally decreases the maximum top displacements, but it would increase the maximum base shear forces, and thus it is necessary to consider the effect of strain rate on the seismic analysis of the transmission tower. The effect of strain rate could be ignored for the seismic analysis of the conductors and ground lines, but the responses of the ground lines considering strain rate effect are larger than those of the conductors. The results could provide a reference for the seismic design of the transmission tower-line system.

INTRODUCTION

Power transmission tower-line system is an important component of a power system. Its reliable operation is of great significance in achieving the energy strategy of nationwide interconnection, power transmission from west to east, and mutual supply between north and south. So far, most of research attentions have been paid on the actions of static load, dynamic characteristic, and nonlinear time history analysis with only geometric nonlinear was taken into account [1–8]. However, several recent cases of damage to transmission towers during earthquakes have proved that the structure under strong earthquake excitation could enter nonlinear status. During the 1976 Tangshan earthquake, some transmission towers collapsed. In the 1999 CHI-CHI earthquake, a lot of transmission lines were broken and some towers collapsed [9]. In 2008, Sichuan electric network was damaged by the Wenchuan earthquake.

Therefore, studies about elastic-plastic analysis of transmission towers have been reported recently. Li et al. [10] calculated the dynamic characters of two different kinds of towers, and the two transmission towers were calculated with two different pieces of seismic records based on the simplified method of seismic calculation, and the nonlinear behaviors and plastic limit of transmission towers in different vibration directions subjected to earthquake action were summarized.

Xiong et al. [11] investigated the elastic-plastic analysis of a long-span concrete filled steel-tube transmission tower under earthquake actions by the static and the dynamic method. The results showed a lot of elements enter the plastic status soon after earthquake action and there were some steel-tube elements that failed but no concrete filled steel-tube element failed at last times. Albermani et al. [12] presented a nonlinear analysis technique for transmission tower structures, and the proposed technique could be used to accurately predict structural failure, with the predictions confirmed by the results of an expensive full-scale test. It is necessary to consider the effect of strain rate on transmission tower-line system using elastic-plastic analysis method. However, there is no research about the analysis of strain rate effect on transmission tower-line system at present.

The experimental results have showed that the steel is a rate sensitive material, which has different mechanical properties under different strain rates. A few experimental studies have been carried out in recent years. ChangandLee [13] investigated the strain rate and strain rate history effects on the inelastic stress-strain behavior of annealed A-36 structural steel at room temperature under monotonic and cyclic loading conditions. Test results show relatively more significant strain rate sensitivity of the material for monotonic loading and less significant strain rate sensitivity for cyclic loading. Restrepo-Posada et al. [14] studied variables affecting cyclic behavior of reinforcing steel, and the effects of strain on the cyclic behavior of reinforcing steel can be very relevant when considering postearthquake retrofitting of reinforced concrete structures. Yang et al. [15] studied the dynamic tensile property of three steels, and the results showed that the yield strength and extension rate of the steel improved with strain rate increasing but the rate sensitivity was different. Song [16] carried out the dynamic cyclic loading test of steel, and the results showed that the yield strength of the steel improved with strain rate increasing, and the elastic modulus was not changed. At present general conclusion is that the yield strength and tensile strength of the steel have a certain improvement with strain rate increasing, and the increased range of yield strength is larger than the tensile strength, but the elastic modulus is not changed. The lower the strength of the steel is, the more obvious the effect of strain rate is. Strain rate has an influence on the response of the structure under earthquake excitation. It is unreasonable to calculate the seismic response of the structure using static analysis, and the effect of strain rate should be considered. It is very important to study the influence of strain rate on the nonlinear seismic response of the structure.

This paper aims to investigate the influence of the strain rate effect on the power transmission tower-line system under different ground motion intensities. A subroutine is developed in ABAQUS. The progressive collapse simulation method is proposed. Incremental nonlinear time history analysis method is adapted to calculate the response with and without considering the strain rate effect. The results could provide a reference for the seismic design of the transmission tower-line system.

STRUCTURAL MODEL

The selected tower for the analysis is illustrated in Figure 1, which has a height of 60.5 m and a square base area of 10.16 m × 10.16 m at ground level. The angle steel with equal section is used for all tower members. Main members of the tower are made of Q345, and secondary members are made of Q235. The mechanical properties of Q345 and Q235 are shown in Table 1.

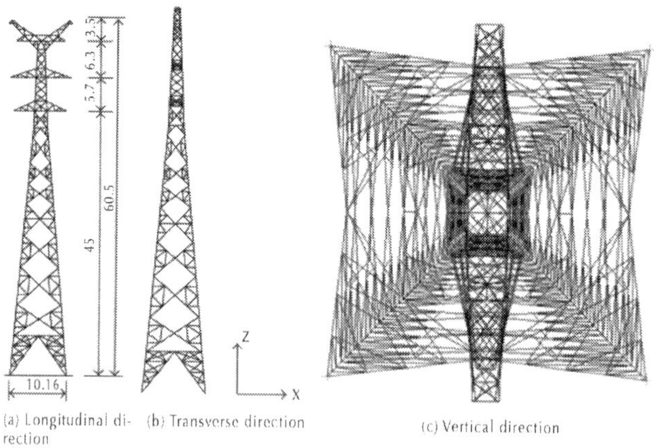

Figure 1: Longitudinal, transverse, and vertical views of a transmission tower (m).

Table 1: Mechanical properties of Q345 and Q235

Material	Yield stress (Pa)	Young's modulus (Pa)	Poisson's ratio	Density (Kg/m^3)
Q345	$3.45E + 08$	$2.00E + 11$	0.3	7800
Q235	$2.35E + 08$	$2.00E + 11$	0.3	7800

A transmission tower-line system consists of many towers and lines, and the coupling effects of tower and line are prominent. However, it is unrealistic to establish a model that includes all towers and lines. The numerical model contains three towers and four-span lines in the paper, which was verified to be reasonable [17, 18]. As shown in Figure 2, the transmission tower-line coupled system includes three towers (1st tower, 2nd tower, and 3rd tower) and four-span line (Span 1, Span 2, Span 3, and Span 4). The upper 8 cables are ground lines and the lower 24 cables are single bundled conductor. The spans of adjacent towers are all 300 m. The conductors and towers are connected with insulators and thc materials for conductor and ground wire are steel-cored aluminum strand. The base points of the transmission tower are fixed on the ground, and the connections between transmission towers and lines are hinged. The side spans of the lines are hinged at the same height of the middle tower. The properties and performance indices of conductor, ground wire, and insulator are listed in Table 2.

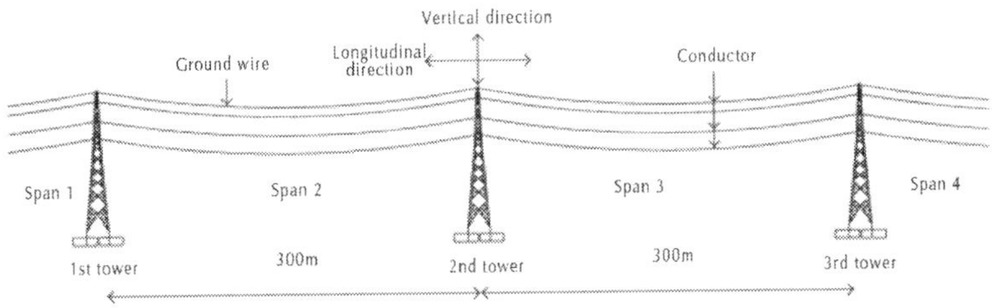

Figure 2: Three-dimensional finite element model of transmission tower-line system.

Table 2: Properties and performance indices of conductor, ground wire, and insulator

Type	Area (m²)	Young's modulus (Pa)	Poisson's ratio	Density (Kg/m³)	Yield force (N)
Conductor	4.25E − 04	6.50E + 10	0.3	3172	98710
Ground wire	1.53E − 04	1.05E + 11	0.3	4631	74200
Insulator	0.02	7.65E + 10	0.2	7500	1000000

The three-dimensional beam elements type, B31, with three translational and three rotational degrees of freedom per node, is employed to model the tower members, and the three-dimensional truss elements type, T3D2, with three translational degrees of freedom per node, is applied to model the lines and insulators in the ABAQUS software. Each tower contains 741 beam elements, each insulator contains one element, and the mesh selected for each conductor and ground wire consists of 100 truss elements.

MATERIAL STRAIN RATE AND ANALYSIS METHOD

Material Strain Rate

The strain rate of steel under seismic excitation is hard to exceed 1/s. The dynamic constitutive relationship model incorporated in the finite element analysis is as follows [19]:

$$f_{yd} = \left(1 + c_f \lg \frac{\dot{\varepsilon}}{\dot{\varepsilon}_0}\right) f_{ys}$$

$$c_f = 0.1709 - 3.289 \times 10^{-4} f_{ys}, \tag{1}$$

where $\dot{\varepsilon}$ is the current strain rate; $\dot{\varepsilon}_0$ is the quasistatic strain rate, $\dot{\varepsilon}_0 = 2.5 \times 10{-}4/s$; f_{ys} is the yield strength at quasistatic strain rate; f_{yd} is the dynamic yield strength at the current strain rate.

Theoretical Analysis

For the transmission tower-line system, the nonlinear dynamic analysis should be used because of its importance and complexity. The motion equation of structure under the seismic excitation is as follows:

$$\begin{aligned} &M(t)\ddot{\chi}(t) + C(t)\dot{\chi}(t) + K(t)\chi(t) \\ &= -M(t)(\ddot{\chi}_u(t) + \ddot{\chi}_v(t) + \ddot{\chi}_w(t)), \end{aligned} \quad (2)$$

where $x^{\cdot\cdot}(t)$, $x^{\cdot}(t)$, and $x(t)$ are the relative acceleration, velocity, and displacement vectors; $x^{\cdot\cdot}u(t)$, $x^{\cdot\cdot}v(t)$, and $x^{\cdot\cdot}w(t)$ are the ground motion accelerations in two horizontal and one vertical directions, respectively; $M(t)$, $K(t)$, and $C(t)$ are the mass, stiffness, and damping matrices, respectively. Generally, $M(t)$ is an invariant, while $K(t)$ and $C(t)$ change during the earthquake.

If the ground motion intensity is weak, the material cannot enter nonlinear status. The mass, stiffness, and damping matrices are irrelevant to strain rate. The strain rate has no influence on the elastic response analysis of the structure. If the ground motion intensity is strong, the material can enter nonlinear status. The stiffness matrix is relevant to strain rate, but the mass matrix is irrelevant to strain rate. The damping matrix is irrelevant to strain rate if the mass-damping is used. However, if stiffness-damping or Raleigh damping is chosen, the damping matrix is relevant to strain rate. Therefore, the strain rate has some influence on the plastic response analysis of the structure.

According to a series of ground motion intensity indices of monotone increasing of a specific seismic wave, the structure is analyzed using nonlinear time history analysis, and the nonlinear seismic response of the structure under different ground motion intensities can be obtained. The method is called incremental dynamic analysis (IDA) [20]. Multiple seismic waves are selected. The response of the structure is calculated by nonlinear time history analysis method with and without strain rate effect taken into account. The peak accelerations of the three-dimensional seismic waves are increased with equal proportions until progressive collapse of the structure occurs.

Progressive Collapse Simulation

Elastic-perfectly plastic material model is used, which is shown in Figure 3. The constitutive relationship is coded using the user subroutine VUMAT [21], which can be implemented in the advanced finite element program ABAQUS. According to the proposed method, once the strain exceeds ultimate strain, the stiffness of the element is zero while the mass remains the same.

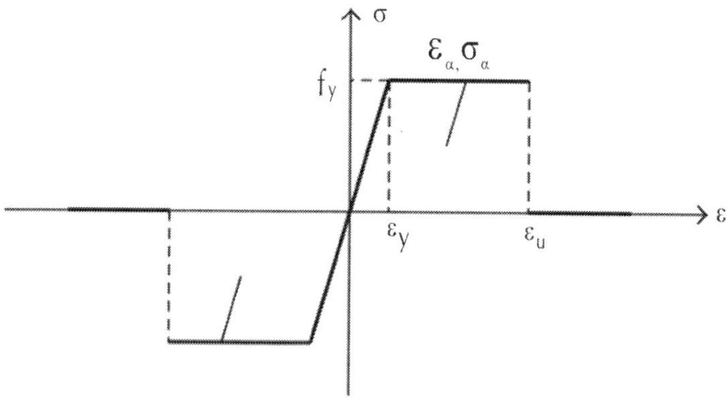

Figure 3: Stress-strain curve of steel.

For uniaxial loading, the stress-strain relationship of steel is as follows:

$$\sigma = \begin{cases} E_s\varepsilon & \varepsilon \le \varepsilon_{y,} \\ f_y & \varepsilon_y \le \varepsilon \le \varepsilon_{u,} \\ 0 & \varepsilon > \varepsilon_{u,} \end{cases} \tag{3}$$

where ε and σ are the strain and stress, respectively; E_s is the elastic modulus; ε_y and f_y are the yield strain and yield stress, respectively; $_u$ is the ultimate strain.

For cyclic loading, the stress-strain relationship is as follows:

$$\sigma = \sigma_a + E_s(\varepsilon - \varepsilon_a), \tag{4}$$

where σ_a and ε_a are the stress and strain of the starting point in the unloading curve, respectively.

THE SELECTION OF SEISMIC WAVES

Three seismic waves are selected: (1) Kobe wave (1995); (2) Northbridge wave (1994); (3) El Centro wave (1940). And the seismic waves include two horizontal components and one vertical component. The NS, WE, and vertical components of seismic wave coincide with the *Y*, *X*, and *Z* directions of the transmission tower-line system shown in Figure 2, respectively. The material of transmission tower under seismic excitation is hard to enter plastic status and is very hard to collapse. When the plastic analysis is carried out, the ground motion intensity requires very strong amplitude. The structure under very strong earthquake action is analyzed using elastic-plastic analysis method. The increasing of ground motion intensity is fast at the beginning of the incremental dynamic analysis. When the structure is close to collapse, the increasing of seismic intensity is slow, and the minimum increment of ground motion intensity is 1 m/s^2.

NUMERICAL SIMULATION AND DISCUSSION

The responses of the strain rate effect on the power transmission tower line system shown in Figure 2 under different ground motion intensities are investigated. The nonlinear time history analysis method is used in the paper. The geometric nonlinearity is taken into account due to large deformation of the transmission lines and the material nonlinearity of the system is considered. The strain rate effect on the maximum top displacement and base shear force of the transmission tower are studied. The strain rate effect on the maximum tension force and vertical displacement of the conductor and ground line are also investigated.

The Effect of Strain Rate on the Transmission Tower

The maximum top displacements of 2nd transmission tower under different seismic excitations are given in Table 3, and the table shows the responses of the transmission tower with and without considering the strain rate. It also can be seen from Table 3 that the transmission tower is easy to collapse under El Centro and Kobe earthquake actions, but it is hard to collapse under Northbridge earthquake action. The deformation capacity of the transmission tower under Kobe earthquake action is larger than those of the other two earthquake actions.

Table 3: Maximum top displacements of 2nd transmission tower

Seismic wave	Peak acceleration (m/s^2)	Displacement (m)			
		X direction		*Y* direction	
		Without considering strain rate	Considering strain rate	Without considering strain rate	Considering strain rate
El Centro	6.0	0.243	0.243	0.330	0.329
	8.0	0.256	0.258	0.453	0.454
	10.0	0.339	0.318	0.514	0.513
	12.0	0.460	0.396	0.618	0.617
	14.0	0.841	0.757	0.691	0.687
	15.0	Collapse	1.124	Collapse	0.706
	16.0	Collapse	Collapse	Collapse	Collapse
Kobe	6.0	0.498	0.501	0.505	0.505
	8.0	0.685	0.682	0.701	0.694
	10.0	1.012	0.973	0.949	0.936
	12.0	1.085	1.020	1.121	1.084
	13.0	1.177	1.109	1.241	1.157
	14.0	Collapse	1.196	Collapse	1.268
	15.0	Collapse	Collapse	Collapse	Collapse

North-bridge	10.0	0.306	0.305	0.329	0.329
	15.0	0.518	0.516	0.453	0.455
	20.0	0.573	0.568	0.499	0.492
	23.0	0.681	0.674	0.644	0.640
	25.0	0.748	0.725	0.664	0.659
	27.0	0.828	0.813	0.743	0.740
	29.0	Collapse	0.880	Collapse	0.752
	30.0	Collapse	Collapse	Collapse	Collapse

Figure 4 shows the relative changes of the maximum top displacements of 2nd transmission tower considering the effect of strain rate with the increasing ground motion intensity. It can be seen form Figure 4 that the effect of strain rate on the transmission tower may decrease or increase the maximum top displacements when the ground motion intensity is weak. When the transmission tower is close to collapse, the transmission tower considering the effect of strain rate significantly decreases the maximum top displacements. The effect of strain rate has little influence on the maximum top displacements of the transmission tower under Northbridge earthquake action. However, the effect of strain rate has a great influence on the maximum top displacements of the transmission tower under El Centro and Kobe earthquake actions, and the maximum reduction of the maximum top displacement is 16.2%. The effect of strain rate on the maximum top displacement of the transmission tower has an increasing tendency with the ground motion intensity increasing.

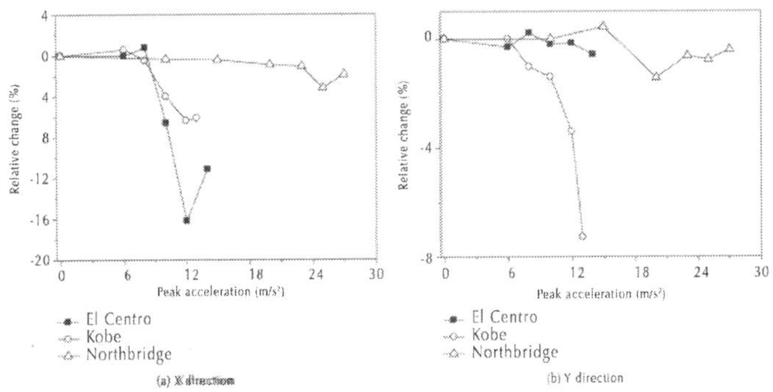

Figure 4: Relative changes of the maximum top displacements of 2nd transmission tower.

The maximum base shear forces of 2nd transmission tower under different seismic wave excitations are shown in Table 4. It can be seen from Table 4 that the base shear forces of the transmission tower under Northbridge earthquake action are very large. Figure 5shows the relative changes of the maximum base shear forces of the transmission tower considering the effect of strain rate with the ground motion intensity increasing. The effect of strain rate on the transmission tower would decrease or increase the maximum base shear forces of the transmission tower. The nonlinear degree of the material is very low and the effect of strain rate is very small when the ground motion intensity is weak. The effect of strain rate would increase the maximum base shear forces of the transmission tower when the seismic ground motion is strong, and the maximum increase is 6.8%. The maximum base shear forces of the transmission tower considering strain rate effect are varied by less than 5%, but the effect of strain rate on the maximum base shear forces of the transmission tower is less than that of the maximum top displacements. The effect of strain rate on maximum base shear forces of the transmission tower has an increasing tendency with the ground motion increasing.

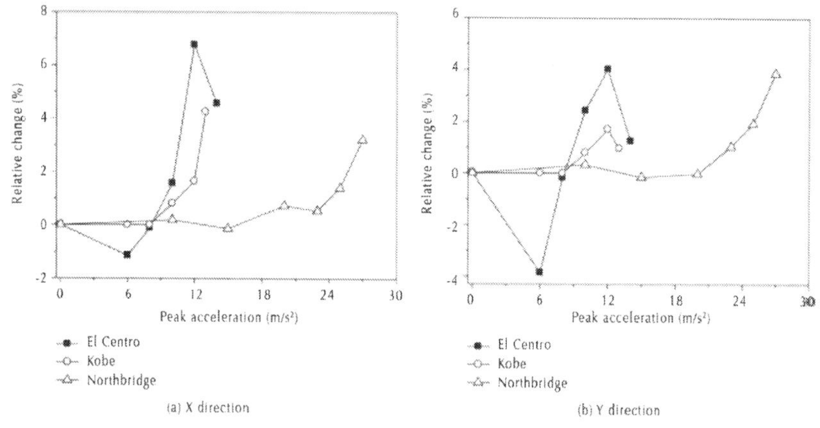

Figure 5: Relative changes of the maximum base shear forces of 2nd transmission tower.

Table 4: Maximum base shear forces of 2nd transmission tower

Seismic wave	Peak acceleration (m/s^2)	Shear force (kN)			
		X direction		*Y* direction	
		Without considering strain rate	Considering strain rate	Without considering strain rate	Considering strain rate
El Centro	6.0	138.465	136.894	163.459	157.404
	8.0	140.797	140.648	214.633	214.248
	10.0	151.760	154.172	262.068	268.623
	12.0	166.223	178.293	303.575	316.353
	14.0	203.141	212.887	319.025	323.093
	15.0	Collapse	222.141	Collapse	307.916
	16.0	Collapse	Collapse	Collapse	Collapse
Kobe	6.0	198.513	198.516	228.912	228.911
	8.0	228.882	228.878	260.429	260.427
	10.0	279.233	281.514	283.561	285.820
	12.0	248.166	252.340	263.036	267.653
	13.0	279.085	291.489	256.623	259.124
	14.0	Collapse	301.570	Collapse	267.427
	15.0	Collapse	Collapse	Collapse	Collapse
North-bridge	10.0	247.764	248.244	216.227	216.941
	15.0	289.418	289.030	309.638	309.181
	20.0	393.213	396.100	453.791	453.690
	23.0	456.353	458.828	469.431	474.372
	25.0	467.501	474.122	464.471	473.617
	27.0	465.700	481.191	500.019	520.178
	29.0	Collapse	485.623	Collapse	523.600
	30.0	Collapse	Collapse	Collapse	Collapse

The Effect of Strain Rate on the Conductor and Ground Line

The maximum tension forces and vertical displacements of the conductor of span 2 are given in Table 5. Relative changes to the conductor's maximum response considering the effect of strain rate with the ground motion increasing are shown in Figure 6. It can be seen from the analysis results that the effect of strain rate has little influence on the response of the conductor, and the maximum responses of the conductor considering the effect of strain rate are varied by less than 1%; so the effect of strain rate on the conductor can be ignored. The material of the conductor cannot enter plastic status, and the effect of strain rate has no direct influence on the response of the conductor, but it could have an indirect influence on the conductor through the transmission tower.

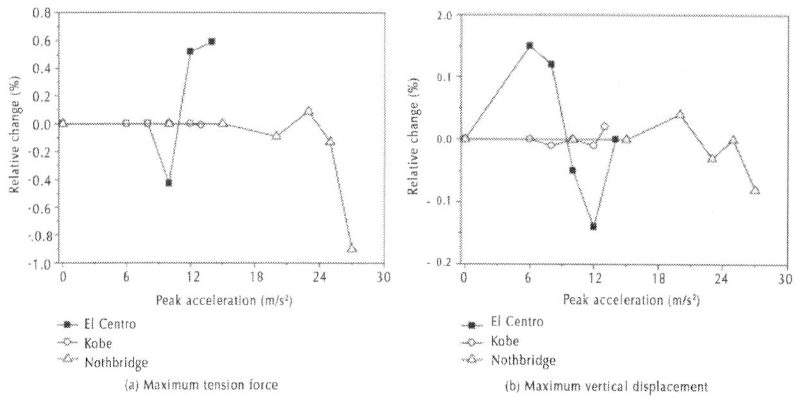

Figure 6: Relative changes of the conductor's maximum responses.

Table 5: Maximum responses of the conductor

Seismic wave	Peak acceleration (m/s²)	Tension force (kN)		Vertical displacement (m)	
		Without considering strain rate	Considering strain rate	Without considering strain rate	Considering strain rate

El Centro	6.0	21.262	21.263	8.859	8.872
	8.0	21.244	21.245	8.625	8.635
	10.0	21.224	21.134	8.423	8.419
	12.0	22.089	22.205	8.420	8.408
	14.0	21.185	21.310	8.184	8.184
	15.0	Collapse	21.651	Collapse	8.197
	16.0	Collapse	Collapse	Collapse	Collapse
Kobe	6.0	21.321	21.321	8.629	8.629
	8.0	21.322	21.322	8.640	8.639
	10.0	21.322	21.323	8.674	8.674
	12.0	21.324	21.323	8.707	8.706
	13.0	21.361	21.359	8.756	8.758
	14.0	Collapse	21.366	Collapse	8.790
	15.0	Collapse	Collapse	Collapse	Collapse
North-bridge	10.0	21.323	21.323	8.377	8.377
	15.0	21.324	21.324	8.081	8.081
	20.0	22.239	22.219	7.386	7.389
	23.0	21.326	21.346	7.219	7.217
	25.0	21.545	21.517	7.220	7.220
	27.0	21.617	21.424	7.221	7.215
	29.0	Collapse	22.203	Collapse	7.223
	30.0	Collapse	Collapse	Collapse	Collapse

Table 6 shows the maximum tension forces and vertical displacements of the ground line of span 2. It can be seen from Table 6 that the change of the response of the ground line is very little with the ground motion increasing. Relative changes of the ground line's maximum responses considering the effect of strain rate with the ground motion increasing are shown in Figure 7. The effect of strain rate has little influence on the ground line response, and the maximum responses of the ground line considering the effect of strain rate are varied by less than 6%. Comparing the response of the conductor and the ground line considering strain rate effect, the effect of strain rate on the ground line is more significant than that of the conductor. The reason is that the insulator length between the ground line and the transmission tower is shorter than that of between the conductor and the transmission tower.

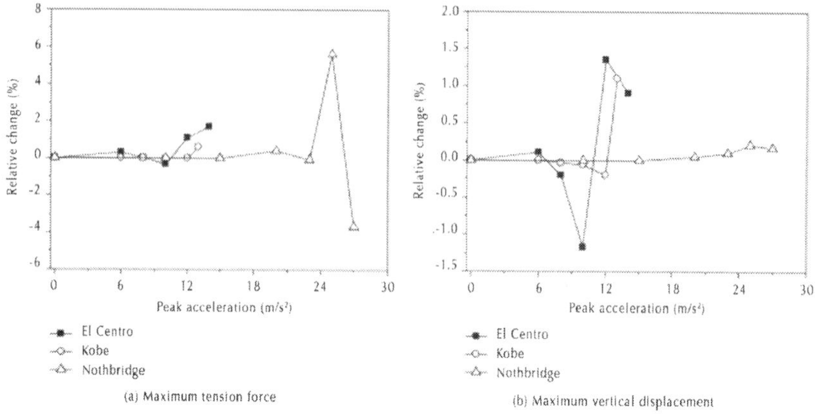

Figure 7: Relative changes of the ground wire's maximum responses.

Table 6: Maximum responses of the ground line

Seismic wave	**Peak acceleration (m/s^2)**	**Tension force (kN)**		**Vertical displacement (m)**	
		Without considering strain rate	**Considering strain rate**	**Without considering strain rate**	**Considering strain rate**
El Centro	6.0	17.239	17.292	3.509	3.513
	8.0	17.847	17.851	3.501	3.494
	10.0	18.651	18.589	4.428	4.377
	12.0	19.282	19.496	4.546	4.608
	14.0	19.183	19.512	4.276	4.315
	15.0	Collapse	19.965	Collapse	4.716
	16.0	Collapse	Collapse	Collapse	Collapse
Kobe	6.0	17.292	17.292	3.484	3.484
	8.0	17.292	17.291	3.485	3.484
	10.0	17.291	17.289	3.614	3.612
	12.0	17.292	17.290	3.586	3.579
	13.0	17.305	17.412	3.968	4.012
	14.0	Collapse	17.460	Collapse	4.163
	15.0	Collapse	Collapse	Collapse	Collapse

North-bridge	10.0	17.296	17.296	5.358	5.358
	15.0	22.424	22.423	5.487	5.487
	20.0	17.539	17.611	5.739	5.742
	23.0	20.882	20.870	5.253	5.258
	25.0	20.115	21.310	5.623	5.635
	27.0	20.790	20.053	5.152	5.161
	29.0	Collapse	21.136	Collapse	7.519
	30.0	Collapse	Collapse	Collapse	Collapse

CONCLUSION

The power transmission tower-line system is analyzed using incremental nonlinear time history analysis method, and the responses of strain rate on the transmission tower-line system are studied. Based on the numerical results, the following conclusions are drawn.

1. If the ground motion is weak, the effect of strain rate on the transmission tower would decrease the maximum top displacements in most of cases, but it would increase the maximum top displacements in few cases. The effect of strain rate on the transmission tower significantly decreases the maximum top displacements when the ground motion intensity of the transmission tower is close to collapse and the maximum reduction of the maximum top displacement is 16.2%.
2. If the ground motion is weak, the effect of strain rate on the transmission tower increases the maximum base shear forces in most of cases, but it would decrease the maximum base shear forces in few cases. The effect of strain rate on the transmission tower would increase the maximum base shear forces when the ground motion intensity of the transmission tower is close to collapse, and the maximum increase of the maximum base shear force is 6.8%.
3. The effects of strain rate on the response of the transmission tower-line system have an increasing tendency with the ground motion intensity increasing.
4. The effect of strain rate has little influence on the response of the conductor, and the maximum responses of the conductor

considering strain rate are varied by less than 1%; so the effect of strain rate on the conductor can be ignored. Comparing the responses of the conductor and ground line considering the effect of strain rate, the effect of strain rate on the ground line is more significant than that of the conductor, but the maximum responses of the ground line considering strain rate effect are less than 6%.

This study demonstrates that the strain rate effect is very important to transmission tower-line system, especially for transmission tower. As many other type towers such as cat head type towers, cup towers, and guyed towers are also widely used in the powers transmission systems, more studies are deemed necessary to further investigate the strain rate effect on responses of these systems.

ACKNOWLEDGMENTS

This work is supported by the National Natural Science Foundation of China under Grant no. 51208285 and China Postdoctoral Science Foundation Funded Project under Grant no. 2012M521338. The support for this research is greatly appreciated.

REFERENCES

1. Li H, Shi W, Wang G, Jia L. Simplified models and experimental verification for coupled transmission tower-line system to seismic excitations. *Journal of Sound and Vibration*. 2005;286(3):569–585.
2. Ghobarah A, Aziz TS, El-Attar M. Response of transmission lines to multiple support excitation. *Engineering Structures*. 1996;18(12):936–946.
3. Li H, Bai H. High-voltage transmission tower-line system subjected to disaster loads.*Progress in Natural Science*. 2006;16(9):899–911.
4. Albermani FGA, Kitipornchai S. Numerical simulation of structural behaviour of transmission towers. *Thin-Walled Structures*. 2003;41(2-3):167–177.

5. Ozono S, Maeda J. In-plane dynamic interaction between a tower and conductors at lower frequencies. *Engineering Structures*. 1992;14(4):210–216.
6. Tian L, Ma R, Li H, Zhang P. Seismic response of straight line type and broken line type transmission lines subjected to non-uniform seismic excitations. *Advanced Steel Construction*. 2014;10(1):85–98.
7. Li H, Bai F, Tian L, Hao H. Response of a transmission tower-line system at a canyon site to spatially varying ground motions. *Journal of Zhejiang University*. 2011;12(2):103–120.
8. Bai F, Hao H, Bi K, Li H. Seismic response analysis of transmission tower-line system on a heterogeneous site to multi-component spatial ground motions. *Advances in Structural Engineering*. 2011;14(3):457–474.
9. Zhang P, Song G, Li H, Lin Y. Seismic control of power transmission tower using pounding TMD. *Journal of Engineering Mechanics*. 2013;139(10):1395–1406.
10. Li H, Hu D, Huang L. Plastic limit analysis of the transmission tower system subjected to earthquake action. *Proceedings of the Chinese Society of Electrical Engineering*. 2006;26(24):192–199.
11. Xiong T, Liang S, Zou L, Wu H. Elastic-plastic analysis of a long-span concrete filled steel-tube transmission tower under earthquake action. *Engineering Mechanics*.2012;29(11):158–164.
12. Albermani F, Kitipornchai S, Chan RWK. Failure analysis of transmission towers.*Engineering Failure Analysis*. 2009;16(6):1922–1928.
13. Chang KC, Lee GC. Strain rate effect on structural steel under cyclic loading. *Journal of Engineering Mechanics*. 1987;113(9):1292–1301.
14. Restrepo-Posada JI, Dodd LL, Park R, Cooke N. Variables affecting cyclic behavior of reinforcing steel. *Journal of Structural Engineering*. 1994;120(11):3178–3196.
15. Yang D, Yu J, Li Z, Duan Z. An experimental study on the tensile properties and damage mechanism of three alloy steel at high strain rate. *ACTA Armamentarii*.1992;3:65–73.

16. Song J. *Investigation of mechanical parameters for rate-sensitive material and its applications [Ph.D. thesis]* Beijing, China: Tsinghua University; 1990. (Chinese)
17. Tian L, Li H, Liu G. Seismic response of power transmission tower-line system subjected to spatially varying ground motions. *Mathematical Problems in Engineering*.2010;2010:20 pages.587317
18. Tian L, Li H, Liu G. Seismic response of power transmission tower-line system under multi-component multi-support excitations. *Journal of Earthquake and Tsunami*.2012;6(4):21 pages.1250025
19. Li M, Li H. Effects of strain rate on reinforced concrete structure under seismic loading. *Advances in Structural Engineering*. 2012;15(3):461–475.
20. ABAQUS. *ABAQUS Analysis User's Manual*. ABAQUS; 2010.
21. Vamvatsikos D, Allin Cornell C. Incremental dynamic analysis. *Earthquake Engineering and Structural Dynamics*. 2002;31(3):491–514.

Chapter 6

Structural Monitoring of Metro Infrastructure during Shield Tunneling Construction

L. Ran,[1] X. W. Ye,[2,*] G. Ming,[1] and X. B. Dong[1]

[1]Hangzhou Metro Group Co., Ltd., Hangzhou 310020, China

[2]Department of Civil Engineering, Zhejiang University, Hangzhou 310058, China

ABSTRACT

Shield tunneling construction of metro infrastructure will continuously disturb the soils. The ground surface will be subjected to uplift or subsidence due to the deep excavation and the extrusion and consolidation of the soils. Implementation of the simultaneous monitoring with the shield tunnel construction will provide an effective reference in controlling the shield driving, while how to design and implement a safe, economic, and effective structural monitoring system for metro infrastructure is of great importance and necessity. This paper presents the general architecture of the shield construction of metro tunnels as well as the procedure of the artificial ground freezing construction of the metro-tunnel cross-passages. The design principles for metro infrastructure monitoring of the shield tunnel intervals in the Hangzhou Metro Line 1 are introduced. The detailed monitoring items and the specified alarming indices for construction monitoring of the shield tunneling are addressed, and the measured settlement variations at different monitoring locations are also presented.

INTRODUCTION

Establishment of the urban metro system will not only efficiently alleviate the problem of urban traffic congestion but also vigorously promote the integrated development among different urban regions [1–3]. Deep excavation and shield tunneling are two critical issues during construction of the metro infrastructure such as metro stations, metro tunnels, and metro-tunnel cross-passages. A significant matter with great concern lies in ensuring the safety of the metro infrastructure at the construction stage, especially when the in-construction metro lines are adjacent to other metro infrastructures or beneath key engineering structures. In the last two decades, a considerable number of civil engineering structures have been instrumented with online structural health monitoring (SHM) systems [4–10]. Investigations pertaining to structural damage detection and safety condition assessment of the concerned structures by use of the long-term measurement data have been conducted by massive researchers worldwide [11–15], while less percent of them have been focused on the underground structures.

In recent years, a significant amount of research has been devoted to the technological progress of the shield tunneling construction and the relevant uncertainty analysis. Goel [16] presented the current status of tunneling and its future potential in India with emphasis on tunneling projects for hydropower developments as well as the tunneling technologies for planning, design, and construction. Boubou et al. [17] analyzed the tunneling-induced ground movement and its correlation with the operation parameters of the tunnel boring machine

(TBM) by use of a nonlinear least square approximation and an artificial neural network (ANN). Špa ková and Straub [18] developed a dynamic Bayesian network (DBN) for probabilistic assessment of tunnel construction performance to facilitate the quantification of uncertainties in the construction process and of the risk from extraordinary events. Fernandez and Moon [19, 20] conducted a theoretical investigation on the significance of the hydromechanical interaction, and the effectiveness of the proposed analytical solution was verified through numerical analysis. Liao et al. [21] studied three typical cases of shield crossings based on the successful engineering projects in Shanghai soft ground and presented the risks of shield crossing sensitive buildings and subways, ground movement prediction and control regulations, and the settings of shield driving parameters.

Regarding the safety monitoring and condition assessment of the underground structures, Bhalla et al. [22] addressed the major technological issues and challenges associated with the structural monitoring of underground structures and presented a detailed review of the available sensor technologies and methods for comprehensive monitoring with special emphasis on conditions encountered underground. Sharma et al. [23] investigated the effect of a large excavation on the deformation of two adjacent mass rapid transit (MRT) tunnels using field monitoring data and finite element analysis. Klar and Linker [24] presented a feasibility study of automated detection of the smuggling tunnels with the development of the smart underground security fence using Brillouin optical time domain reflectometry (BOTDR). Shao and Macari [25] systematically introduced an effective and reliable method in predicting the deformation for the deep excavation, relying on the accurate field-measured displacement as the input information and feedback to a numerical model of the deep excavation. Kavvadas [26] described the types of ground deformation measurements often used in tunneling, the difficulties in obtaining ground deformation measurements, the methods commonly used for evaluation, and the application of these measurements in establishing early warning systems. Bennett et al. [27] introduced the challenges involved with a wireless sensor network (WSN) installation in an underground environment and described two case studies of field trials of WSN systems.

DESCRIPTION OF METRO-TUNNEL SHIELD SECTIONS

Scope of Shield Tunnel Intervals

As illustrated in Figure 1, the studied shield tunnel intervals, namely, Genshanmen Station-Wenhua Square Station interval and Wenhua Square Station-Wulin Square Station interval, are two important shield tunneling construction sections in Hangzhou Metro Line 1, which is the first metro line of the urban rapid rail transit system in Hangzhou, China. For the Genshanmen Station-Wenhua Square Station interval, the shield tunnel extends from the Wenhua Square Station and arrives at Wenhui Road after turning right from the Zhongshan North Road. This shield tunnel line will travel beneath 13 residential buildings of Zhaohui community as well as the bridge piers of the Shangtang Viaduct, and part of the shield tunnel also will pass beneath two rivers. As per the Wenhua Square Station-Wulin Square Station interval, the shield tunnel starts from the east of the Wulin Square Station and travels beneath Huancheng North Road and the Beijing-Hangzhou Grand Canal. It gradually enters Zhongshan North Road from the West Lake Wenhua Square and then reaches the Wenhua Square Station being as two overlapping tunnels.

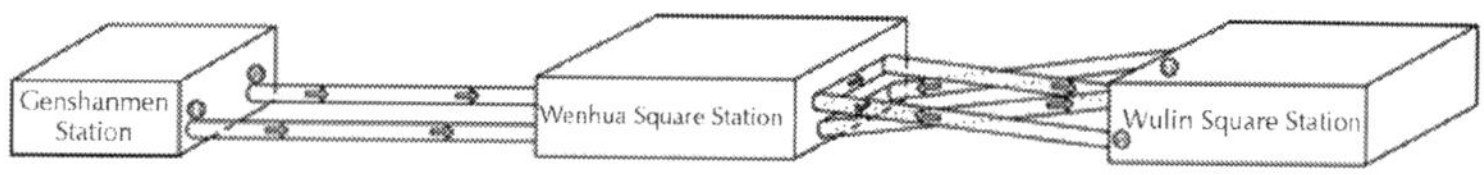

Figure 1: Layout of tunneling construction of metro shield intervals.

Geological Conditions

The geological structure of the shield tunnel section consists of a considerable amount of silty soils and sandy soils. Due to the differences of accumulation periods and consolidation conditions, the compaction extent varies from loose condition to medium density in the vertical direction. The properties of the soil layers are different, and the general thickness of the soil layers is about 20 m. The silt soft soils and clayey soils are buried beneath with a dense condition and the thickness of the soil layers is about 40~45 m. The depth of the bottom bedrock is about 55~63 m from the ground surface. According to the geological properties and forming conditions, the foundation soils are divided into different soil layers, as listed in Table 1 in detail. The shield tunnels of the Genshanmen Station-Wenhua Square Station interval mainly pass through the soil layers 4_4, 6_1, 6_2, 7_1, 8_2, and 9_1, while the shield tunnels of the Wenhua Square Station-Wulin Square Station interval go through the soil layers 4_2, 4_3, 6_1, 7_1, and 7_2.

Table 1: Property description of soil layers in shield tunnel section

Number	Soil layer	Detail description
1	Miscellaneous fill soil	Mainly composed by construction and domestic waste (0.4~3.0 m)
2	Plain fill soil	Generally composed by the silty soil with organic matters (0.4~3.4 m)
3	Silty soil	Containing lots of humic substances and scrap micas (0.6 m)
4	Silty clay	With much iron oxides and scrap micas (0.6~1.8 m)
5	Clayey silt	With much iron oxides and scrap micas (0.3~2.2 m)
6	Sandy silt	With much iron oxides and scrap micas
7	Clayey silt	With organic matters and scrap micas
8	Clayey silt	With organic matters and scrap micas
9	Sandy silt	With organic matters and scrap micas
10	Clayey silt	With organic matters and scrap micas
11	Sandy silt	With organic matters and scrap micas
12	Sandy silt	With organic matters, iron oxides, and scrap micas
13	Silty clay	With organic matters and scrap micas
14	Silty soft clay	With organic matters and silty soils (2.3~6.8 m)

15	Silty soil	With organic matters and sludge soils (2.3~7.0 m)
16	Silty soil	With organic matters (3.4~7.9 m)
17	Clayey silt	With organic matters, scrap micas, and silty clay
18	Silty soil	With organic matters (3.7~9.4 m)
19	Silty soil	With organic matters (2.1~6.6 m)
20	Clay	With organic matters and silty soils (2.3~6.9 m)
21	Clay	With organic matters and silty soils
22	Gray clay	With organic matters and silts (2.0~10.4 m)
23	Silty clay	With organic matters and scrap micas (1.6~5.2 m)
24	Silty clay	With organic matters and scrap micas (0.6~4.8 m)

ORGANIZATION OF SHIELD TUNNEL CONSTRUCTION

General Planning of Shield Tunneling

After a comprehensive investigation on the type and characteristics of the metro-tunnel shield sections, various construction teams are established, namely, shield construction team, tunnel segment construction team, and integration construction team. During the whole construction process, the shield construction should be prepared in advance and the accessory structures will be constructed during the construction intermittence of the main structures. The construction time nodes are settled in accordance with the completion of the construction of the milestone part of the project. The timely completion of the construction of the shield tunnel intervals is ensured through harmonious construction planning. By doing so, the general organization of the shield tunnel construction will be legitimately controlled.

As illustrated in Figure 2, two shield tunneling machines are used in the concerned shield tunnel intervals. The 10th shield machine starts from the left line at the Genshanmen Station and will be removed from the Wenhua Square Station. It will be moved again to the right line of the Wulin Square Station and reaches the Wenhua Square Station and then ends at Wulin Square Station. The 11th shield machine starts from the right line at the Genshanmen Station and is removed from the Wenhua Square Station. It will be moved again to the left line of the Wulin Square Station and arrives at the Wenhua Square Station and then ends at Wulin Square Station. Figure 3 shows the site photo of the shield tunnel construction.

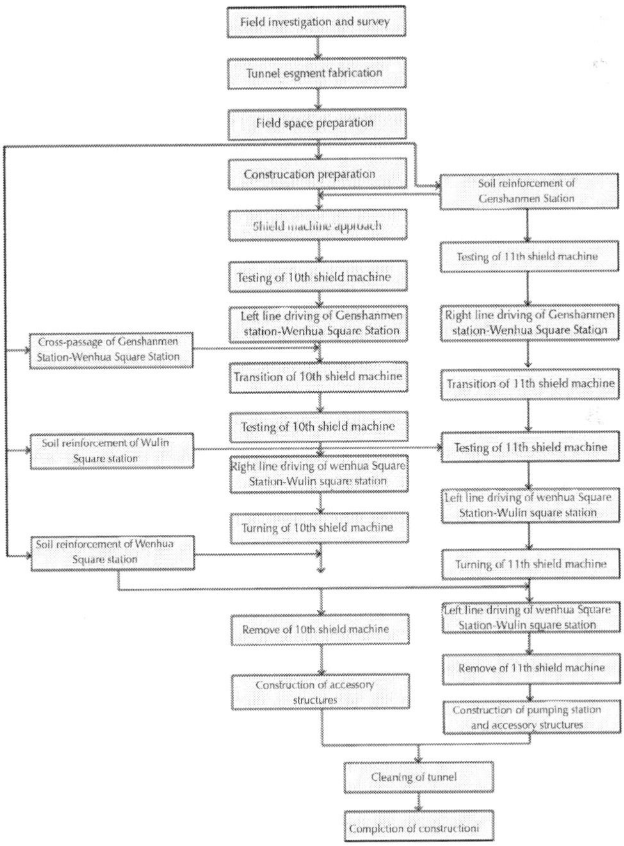

Figure 2: Framework of shield tunnel construction.

Figure 3: Photo of field shield tunnel construction.

Freezing Construction of Cross-Passages

Following the Chinese code of metro design, a cross-passage between two tunnels should be set up if the continuous length of one of the tunnels is larger than 600 m. On the basis of the geological conditions of the shield tunnel sections, the soils around the tunnels are reinforced through the horizontal freezing and then the tunnel linings are constructed after excavation. By doing so, a high-strength sealing frozen curtain will be formed to guarantee the safety of excavation construction. Figure 4 shows the detailed deployment of the freezing tubes during the artificial ground freezing construction. The procedure of the artificial ground freezing construction is illustrated in Figure 5.

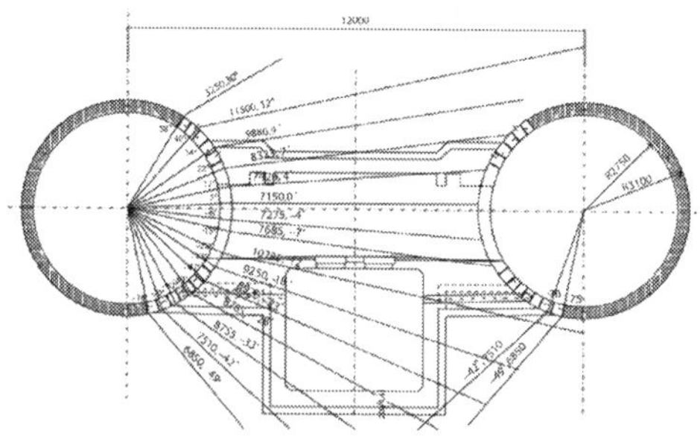

Figure 4: Arrangement of freezing tubes.

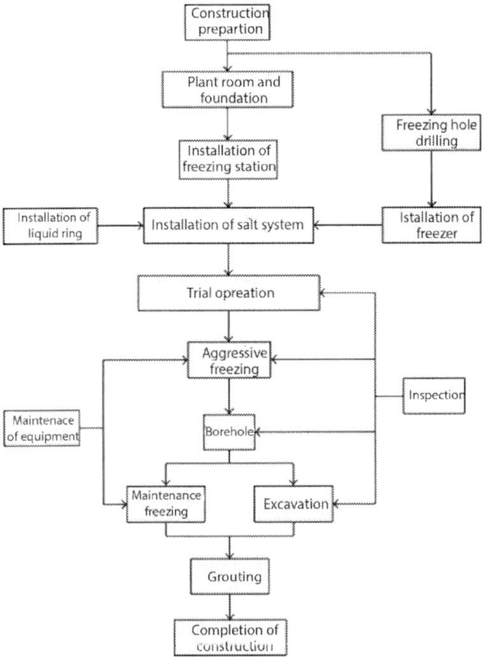

Figure 5: Procedure of artificial ground freezing construction.

IN-CONSTRUCTION MONITORING AND SAFETY EVALUATION

Design Principles for Metro Infrastructure Monitoring

For the managers of metro infrastructure, the implementation of efficient surveillance and measurement of engineering projects with large risks is crucial such as the deep excavations with a depth over 5 m and the shield tunnel intervals. The execution of the informationalised construction through in-construction monitoring can make the engineering measurement normal and standard. The design principles for monitoring the metro infrastructure include the following: (i) all the monitoring items are effectively synthesized and the measurement data can be interactively verified; (ii) a comprehensive monitoring scope should be established to ensure that the measurement data are promptly acquired with a continuous and accurate manner; (iii) the mature and advanced measurement equipment is adopted and should be protected during installation; (iv) the critical parameters in the structural design are monitored for optimization; (v) special attention should be paid to the key construction sections for dense monitoring; (vi) the monitoring types, locations, and protection measures are determined based on the realistic construction; and (vii) the monitoring method is selected in accordance with the rules of concision and cost-effectiveness. Figure 6 illustrates the flowchart of the systematical surveillance and measurement during metro-tunnel shield construction.

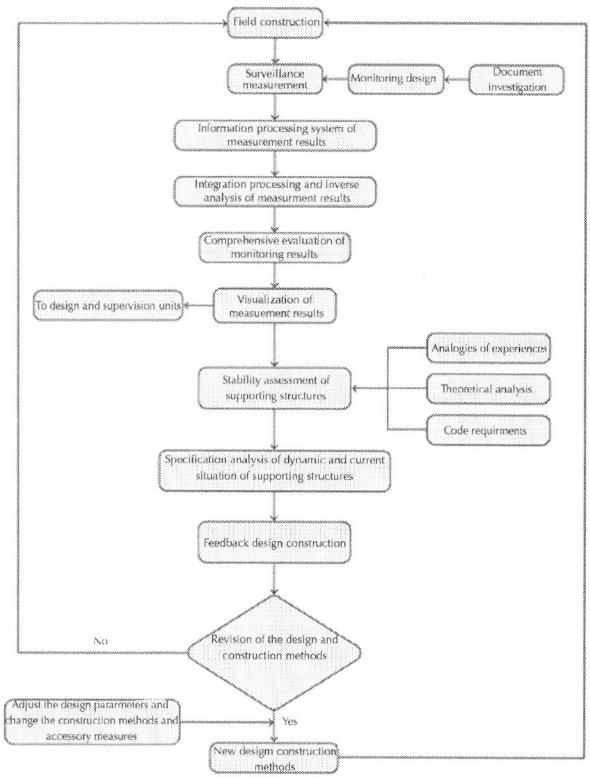

Figure 6: Flowchart of systematical surveillance and measurement.

Monitoring Items and Result Analysis

In this project, the monitoring items include the settlement, inclination and cracking of the surrounding buildings, the vertical displacement of the adjacent underground colligation pipelines, the settlement of the ground surface, the underground water level, the sink deformation of the tunnel vault, the tunnel bottom heave, and the tunnel clearance convergence. The alarming indices are generally stipulated based on the total variations and the variation velocities of the monitoring parameters, and the warning value of the total variation should not be larger than the design value. Table 2 lists the specified alarming indices for monitoring of the shield tunnel construction. Figure 7 shows the measured settlement variations at different monitoring locations of the uplink and downlink tunnels at the Wenhua Square Station-Wulin Square Station interval.

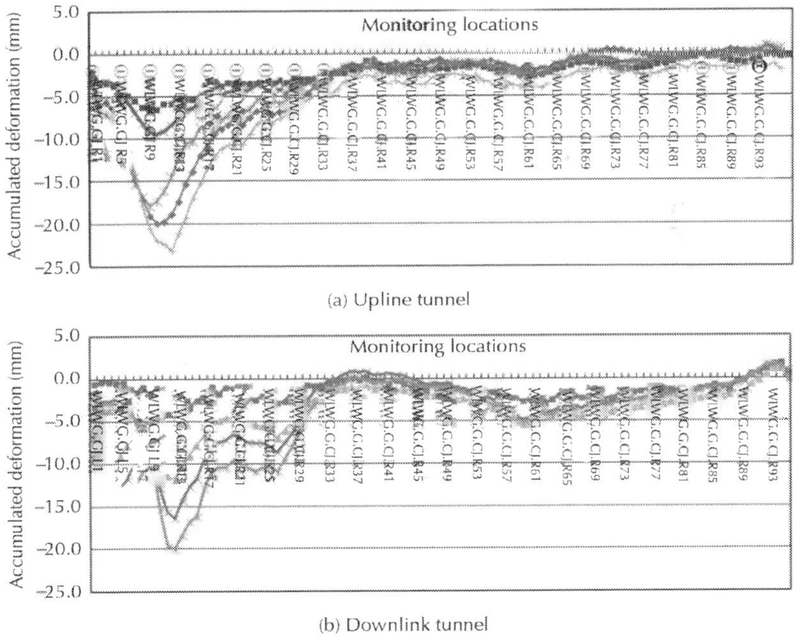

Figure 7: Monitored settlement variations at different monitoring locations.

Table 2: Alarming indices for construction monitoring of shield tunneling

Item	Alarming index		
	Daily variation	**Design value**	**Warning value**
Deformation of adjacent underground colligation pipelines	2 mm	20 mm	80% of design value
Settlement of adjacent ground surface	3 mm	+10 mm/−30 mm	
Vertical displacement of surrounding buildings	2 mm	20 mm	
Settlement of tunnel	2 mm	20 mm	
Tunnel clearance convergence	2 mm	30 mm	

CONCLUSIONS

Development and utilization of the urban underground space have gained increasing concerns from the urban administrative agency and the civil engineering community, which has been a significant flag for urban modernization. In the past years, the design and construction of the urban underground structures are considered in terms of the strength and stability of the structural components. However, the deformation control design is deemed to be dominant for the deep excavation of the metro infrastructure due to the increasing number of the surrounding buildings as well as the scope and depth of the metro-tunnel shield construction. In this connection, monitoring of the metro infrastructure during the shield tunnel construction becomes important which will facilitate the condition tracking and safety evaluation of the metro infrastructure with the help of the long-term measurement data.

ACKNOWLEDGMENTS

The work described in this paper was jointly supported by the National Science Foundation of China (Grant nos. 51308493 and U1234204), the Research Fund for the Doctoral Program of Higher Education of

China (Grant no. 20130101120080), and the Fundamental Research Funds for the Central Universities of China (Grant no. 2013QNA4023).

REFERENCES

1. Ran L, Yi TH, Ye XW, Dong XB. Long-term deformation monitoring of metro-tunnel airshaft excavation during construction stage. *International Journal of Distributed Sensor Networks*. 2012;2012:11 pages.972893
2. Ye XW, Ni YQ, Yin JH. Safety monitoring of railway tunnel construction using FBG sensing technology. *Advances in Structural Engineering*. 2013;16(8):1401–1409.
3. Ye XW, Ran L, Yi TH, Dong XB. Intelligent risk assessment for dewatering of metro-tunnel deep excavations. *Mathematical Problems in Engineering*. 2012;2012:13 pages.618979
4. Chen B, Xu YL, Zhao X. Integrated vibration control and health monitoring of building structures: a time-domain approach. *Smart Structures and Systems*.2010;6(7):811–833.
5. Li S, Li H, Liu Y, Lan C, Zhou W, Ou J. SMC structural health monitoring benchmark problem using monitored data from an actual cable-stayed bridge. *Structural Control and Health Monitoring*. 2014;21(2):156–172.
6. YiT-H, LiH-N, GuM. Optimal sensor placement for structural health monitoring based on multiple optimization strategies. *Structural Design of Tall and Special Buildings*.2011;20(7):881–900.
7. Yi T-H, Li H-N, Gu M. Recent research and applications of GPS-based monitoring technology for high-rise structures. *Structural Control and Health Monitoring*.2013;20(5):649–670.
8. Ye XW, Ni YQ, Wai TT, Wong KY, Zhang XM, Xu F. A vision-based system for dynamic displacement measurement of long-span bridges: algorithm and verification.*Smart Structures and Systems*. 2013;12(3-4):363–379.
9. Ying YJ, Garrett Jr JH, Oppenheim IJ, et al. Toward data-driven structural health monitoring: application of machine learning and signal processing to damage detection.*Journal of Computing in Civil Engineering*. 2013;27(6):667–680.

10. Ni YQ, Ye XW, Ko JM. Modeling of stress spectrum using long-term monitoring data and finite mixture distributions. *Journal of Engineering Mechanics*. 2011;138(2):175–183.

11. Chen B, Sun Y-Z, Wang G-J, Duan L-Y. Assessment on time-varying thermal loading of engineering structures based on a new solar radiation model. *Mathematical Problems in Engineering*. 2014;2014:15 pages.639867

12. Tibaduiza D-A, Torres-Arredondo M-A, Mujica LE, Rodellar J, Fritzen C-P. A study of two unsupervised data driven statistical methodologies for detecting and classifying damages in structural health monitoring. *Mechanical Systems and Signal Processing*.2013;41(1-2):467–484.

13. Catbas FN, Gul M, Gokce HB, Zaurin R, Frangopol DM, Grimmelsam KA. Critical issues, condition assessment and monitoring of heavy movable structures: emphasis on movable bridges. *Structure and Infrastructure Engineering*. 2014;10(2):261–276.

14. Ye XW, Ni YQ, Wong KY, Ko JM. Statistical analysis of stress spectra for fatigue life assessment of steel bridges with structural health monitoring data. *Engineering Structures*. 2012;45:166–176.

15. Ni YQ, Ye XW, Ko JM. Monitoring-based fatigue reliability assessment of steel bridges: analytical model and application. *Journal of Structural Engineering*.2010;136(12):1563–1573.

16. Goel RK. Status of tunnelling and underground construction activities and technologies in India. *Tunnelling and Underground Space Technology*. 2001;16(2):63–75.

17. Boubou R, Emeriault F, Kastner R. Artificial neural network application for the prediction of ground surface movements induced by shield tunnelling. *Canadian Geotechnical Journal*. 2010;47(11):1214–1233.

18. Špa ková O, Straub D. Dynamic Bayesian network for probabilistic modeling of tunnel excavation process. *Computer-Aided Civil and Infrastructure Engineering*.2013;28(1):1–21.

19. Fernandez G, Moon J. Excavation-induced hydraulic conductivity reduction around a tunnel—part 1: guideline for estimate of

ground water inflow rate. *Tunnelling and Underground Space Technology*. 2010;25(5):560–566.

20. Fernandez G, Moon J. Excavation-induced hydraulic conductivity reduction around a tunnel—part 2: verification of proposed method using numerical modeling. *Tunnelling and Underground Space Technology*. 2010;25(5):567–574.
21. Liao S-M, Liu J-H, Wang R-L, Li Z-M. Shield tunneling and environment protection in Shanghai soft ground. *Tunnelling and Underground Space Technology*.2009;24(4):454–465.
22. Bhalla S, Yang YW, Zhao J, Soh CK. Structural health monitoring of underground facilities—technological issues and challenges. *Tunnelling and Underground Space Technology*. 2005;20(5):487–500.
23. Sharma JS, Hefny AM, Zhao J, Chan CW. Effect of large excavation on deformation of adjacent MRT tunnels. *Tunnelling and Underground Space Technology*.2001;16(2):93–98.
24. Klar A, Linker R. Feasibility study of automated detection of tunnel excavation by Brillouin optical time domain reflectometry. *Tunnelling and Underground Space Technology*. 2010;25(5):575–586.
25. Shao Y, Macari EJ. Information feedback analysis in deep excavations. *International Journal of Geomechanics*. 2008;8(1):91–103.
26. Kavvadas MJ. Monitoring ground deformation in tunnelling: current practice in transportation tunnels. *Engineering Geology*. 2005;79(1-2):93–113.
27. Bennett PJ, Soga K, Wassell I, et al. Wireless sensor networks for underground railway applications: case studies in Prague and London. *Smart Structures and Systems*.2010;6(5-6):619–639.

Chapter 7

A Microbial Fuel Cell in Contaminated Ground Delineated By Electrical Self-Potential and Normalized Induced Polarization Data

L. Ran,[1] X. W. Ye,[2,*] G. Ming,[1] and X. B. Dong[1]

R. Doherty[1], B. Kulessa[2,6], A. S. Ferguson[3,6], M. J. Larkin[4], L. A. Kulakov[4] and R. M. Kalin[5]

[1]Environmental Engineering Research Centre, School of Planning Architecture and Civil Engineering, Queen's University of Belfast, Belfast, UK

[2]School of the Environment and Society, Swansea University, Swansea, UK

[3]Department of Civil Engineering and Engineering Mechanics, Columbia University, New York, New York, USA

[4]QUESTOR Centre and School of Biological Sciences, Queen's University of Belfast, Belfast, UK

[5]David Livingstone Centre for Sustainability, Department of Civil Engineering, Strathclyde University, Glasgow, UK

[6]Environmental Engineering Research Centre, School of Planning Architecture and Civil Engineering, Queen's University of Belfast, Belfast, UK.

ABSTRACT

There is a growing interest in the use of geophysical methods to aid investigation and monitoring of complex biogeochemical environments, for example delineation of contaminants and microbial activity related to land contamination. We combined geophysical monitoring with chemical and microbiological analysis to create a conceptual biogeochemical model of processes around a contaminant plume within a manufactured gas plant site. Self-potential, induced polarization and electrical resistivity techniques were used to monitor the plume. We propose that an exceptionally strong (>800 mV peak to peak) dipolar SP anomaly represents a microbial fuel cell operating in the subsurface. The electromagnetic and electrical geophysical data delineated a shallow aerobic perched water body containing conductive gasworks waste which acts as the abiotic cathode of microbial fuel cell. This is separated from the plume below by a thin clay layer across the site. Microbiological evidence suggests that degradation of organic contaminants in the plume is dominated by the presence of ammonium and its subsequent degradation. We propose that the degradation of contaminants by microbial communities at the edge of the plume provides a source of electrons and acts as the anode of the fuel cell. We hypothesize that ions and electrons are transferred through the clay layer that was punctured during the trial pitting phase of the investigation. This is inferred to act as an electronic conductor connecting the biologically mediated anode to the abiotic cathode. Integrated electrical geophysical techniques appear well suited to act as rapid, low cost sustainable tools to monitor biodegradation.

INTRODUCTION

The use of sustainable remediation methods, such as permeable reactive barriers, at complex contaminated sites requires an insightful and multidisciplinary approach [*Kalin*, 2004; *Gibert et al.*, 2007].

The regulatory level of detail required to monitor permeable reactive barriers, coupled with the elevated costs of in situ and laboratory analyses can quickly render this environmentally friendly approach to brownfield risk management and remediation unsustainable. The need for novel, low cost, low impact and sustainable remediation methods and monitoring tools are critical if brownfield redevelopment issues are to be addressed economically [*Spira et al.*, 2006]. This paper applies electrical geophysical methods such as self-potential, induced polarization and resistivity to a site with a contaminant plume that is managed by a permeable reactive barrier. Electrical geophysical methods have been applied as an investigation and monitoring method to determine a variety of environmental conditions. The self-potential method measures the electrical potential that arises from natural current flow in the subsurface, which is often due to complex and non-unique mechanisms. A principal source mechanism is subsurface fluid flow as an electrical charge separation occurring between the solid matrix and the pore fluid (electrical streaming potential). This has identified piezometric surfaces utilizing the electrokinetic properties of water rock interaction [*Fournier*, 1989; *Revil et al.*, 2003; *Linde et al.*, 2007]. Another principal self-potential mechanism is due to electrochemical processes resulting in the diffusion of ions. The self-potential electrochemical potentials arising from chemical gradients has been used to monitor groundwater tracer tests [*Sandberg et al.*, 2002] and to define mixing of glacial meltwaters [*Kulessa et al.*, 2003]. Redox driven electrochemical gradients have produced self-potential signals associated with ore deposits [*Sato and Mooney*, 1960; *Bigalke and Grabner*, 1997] which coined the term 'geobattery' and more recently with contaminant plumes [*Naudet et al.*, 2004, *Arora et al.*, 2007; *Minsley et al.*, 2007]. Laboratory work has helped to define the relationship between self-potential and redox potential [*Maineult et al.*, 2006; *Castermant et al.*, 2008; *Revil et al.*, 2009] where the self-potential should be constrained by the range of Eh and should require the presence of an electronic conductor. In addition to these traditionally recognized self-potential sources, microbially mediated coupled ion-electron flows have recently emerged as a subset of the electrochemical mechanism for natural subsurface current flow [*Revil et al.*, 2010]. Microbially mediated redox environments may produce electronic conductors through the bioprecipitation of minerals allowing bio-geobatteries to occur [*Naudet and Revil.*, 2005]. The electrical resistivity and induced polarization methods respectively exploit the resistance of the subsurface to the flow of an injected current, and the ability of the subsurface to store electrical charge [*Reynolds*, 1997; *Sharma*, 1997]. Laboratory experiments using induced polarization have identified biomineralization [*Slater et al.*, 2007; *Personna et al.*, 2008] microbial presence [*Ntarlagiannis et al.*, 2005;*Davis et al.*, 2006] and artificial biofilms [*Ntarlagiannis and Ferguson*, 2009]. Field applications of induced polarization with respect to contamination issues have focused on monitoring of remediation performance [*Slater and Binley*, 2006; *Williams et al.*, 2009]. This work compiles investigation and monitoring using electrical geophysical methods at a former manufactured gas plant site to create a conceptual biogeochemical model of processes associated with a contaminant plume. We present a field-scale case study that considers self-potential, induced polarization and electrical resistivity applied to characterize a complex contaminant plume. More specifically, the electrical geophysical data are used to update the biogeochemical model of the complex contaminant plume, and propose a microbial fuel cell operating in the subsurface that is an indicator of the biodegradation process.

SITE DESCRIPTION

The site is a former manufactured gas plant at Portadown, Northern Ireland; it occupies an area of approximately 1 ha and had been operational for over one hundred years. The west of the site contained the foundations of demolished structures such as the gas-holding tanks and tar well (Figure 1). The eastern sector of the site contained dumped gasworks waste in the north with undisturbed alluvial sediments in the south. During 1999, a series of multi disciplinary investigation work was undertaken to assess the level of subsurface contamination [*Ferguson et al.*, 2003]. A risk management strategy (a permeable reactive barrier) was implemented during 2001 and monitoring at the site has continued to the present. A time line of investigation, risk management and monitoring at the site is outlined in Table 1.

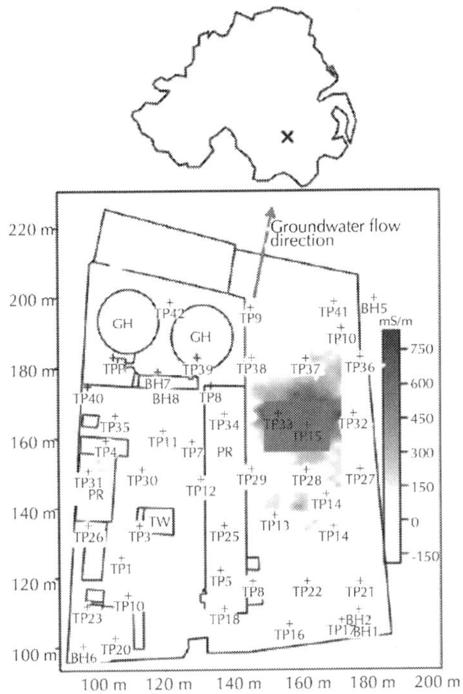

Figure 1: Location (cross) and site plan of the former manufactured gas plant at Portadown, Northern Ireland, with sample locations on a local grid using an arbitrary datum. Labels of foundations of demolished infrastructure prior to remediation are as follows: GH = gas-holding tanks, TW = Tar Well, PR = Purifiers, TP = Trial Pit, BH = Borehole. The reconnaissance EM31 apparent conductivity data (in mS/m) are characterized by a strongly conductive anomaly in the eastern sector of the site.

Table 1: Timeline of Investigation, Remediation, and Monitoring at the Manufactured Gas Plant Site, Portadown, Northern Ireland

Date	Event
1880s–1980s	Operational life of the Manufactured Gas Plant (Gasworks)
March 1999	First geophysical survey using EM31, EM38, and EM61
May–July 1999	Intrusive site investigation 43 trial pits and 8 boreholes
June 2001–October 2002	Installation of cement bentonite slurry wall and Permeable reactive barrier treatment zone

October 2002 - present	Monitoring of permeable reactive barrier treatment zone
March 2005	Geophysics survey of contaminant plume using self-potential, induced polarization and resistivity

METHODOLOGY

Reconnaissance EM Geophysical Surveys

Reconnaissance electromagnetic geophysical surveys on a 10 m × 10 m grid, using three different GEONICS EM instruments, provided direction for the intrusive site investigation. As described in most textbooks of environmental geophysics [e.g., *Reynolds*, 1997; *Sharma*, 1997], electromagnetic geophysical methods consider the generation of a primary electromagnetic field at the ground surface using one coil in the survey instrument. The electromagnetic response of the subsurface to that primary field is measured with a second instrument coil. Comparison of the signal characteristics of this secondary electromagnetic field with the primary field allows conductive materials in the subsurface to be detected and laterally delineated. The underlying physical principle focuses on the secondary electromagnetic field being generated by eddy currents induced by the primary field in conductive subsurface materials. Thus, the more conductive the subsurface material, the stronger the eddy currents and the stronger the secondary electromagnetic field. The depth of penetration and spatial resolution depends on the electromagnetic instrument utilized. In the present case GEONICS EM-61, EM-38 and EM-31 were used, designed respectively as a metal detector, a soil salinity probe and a bulk ground-conductivity sensor.

Intrusive Site Investigation and Chemical and Microbial Sampling and Analyses

The intrusive site investigation was carried out in two phases. The first phase involved forty three trial pits to a maximum depth of 5m with soil and groundwater samples recovered at regular intervals for contaminant chemistry and microbial analysis. The contaminant chemistry analysis for soil and water samples from trial pit was carried out at an accredited laboratory (Geochem, Chester, UK). The analysis suite included inorganic contaminants (ammonium, cyanide, sulphur, and sulphate), metals (As, Cr, Cu, Ni, Pb, Se, Zn, and Hg) and organic contaminants (solvent extractable matter, mineral oil, non-volatile aromatics, and resins). Based on the trial pit contaminant chemistry results, eight boreholes were drilled and emplaced with multi level piezometers to be used in conjunction with the risk management strategy adopted at the site. Microbial analysis from the trial pit samples included enumeration of microorganisms, cloning 16S rRNA genes and sequencing. Total viable counts of aerobic heterotrophic bacteria were enumerated and bacteria were isolated using R2A agar plates (OXOID Ltd.) We also extracted DNA for Polymerase chain reaction amplification. DNA extraction used the FastDNA SPIN Kit for Soil (BIO 101) and followed the manufacturer's protocol. DNA samples were chronologically logged and stored for future reference. After DNA extraction from soil the 16S rDNA genes were amplified by PCR (primers 8F and 518R from [*Leu et al.*, 1998]). The ~500 kb DNA fragments were purified, cloned into pUC129, and transferred into *E. coli* DH5a [*Sambrook et al.*, 1989]. The clones (over 100) were then analyzed by restriction analysis using three restriction enzymes (*Alu*I, *Rsa*I, and *Hpa*II) to eliminate identical clones. All the clones showing a different restriction pattern were sequenced in-house using the Beckman CEQ2000 automated sequencer. For the purpose of the initial phylogenetic identification, obtained sequences were compared to GenBank entries using BLAST engine (http://blast.ncbi.nlm.nih.gov/Blast.cgi).

Conceptual Contaminant Transport Model

The data from the site investigation were used to create a conceptual and numerical groundwater-flow model of the site. The model was calibrated and verified during a dewatering phase to remove the gas holding tank foundations (Figure 1) prior to cement bentonite slurry wall construction [*Doherty et al.*, 2003]. A simple contaminant transport model was constructed using the tar well on the site as a continuous source of 500 mg/L of phenanthrene. This well had been recorded at the site from its inception, and was inferred to be the main source of groundwater contamination based on the synthesized post-investigation model of the site. Phenanthrene was chosen as representative organic compound migrating in groundwater owing to its recalcitrant nature in terms of mobility and solubility. Other potential sources of contamination, such as from buildings or dumped gasworks waste, were not considered because the duration of a continuous contaminant source could not be established. The contaminant model was run for a period of one hundred years to mimic the transport of the more recalcitrant polycyclic aromatic hydrocarbons over the history of the site.

Electrical Geophysical Surveys in Support of Remediation Monitoring

The longer-term risk management strategy of the soil and groundwater contamination focused on the installation of a permeable reactive barrier at the down flow boundary of the site, along with a cement bentonite slurry wall for groundwater management through the permeable reactive barrier. During March 2005 as part of the remediation monitoring process, self-potential data were collected at a total of 110 measurement stations on a 5m × 5m grid, covering an area of 45 × 50 m (2250m^2). These surveys were complemented by ten parallel induced polarization and electrical resistivity profiles that were co-located with the self-potential grid locations in an East-to-West direction. The electrical geophysical survey area focused on the central portion of the site covering pristine natural ground, known contamination and anomalies identified by the reconnaissance electromagnetic geophysical surveys.

In the absence of significant thermally generated self-potentials, we expect our raw self-potential data to be an integrated signature of streaming, electrochemical and biologically mediated electrical potentials. We used off-the-shelf non-polarizing Pb-PbCl electrodes [*Petiau*, 2000] for our self-potential surveys, together with a METRA HIT 22S high-impedance multimeter and rugged single-core wire. To ensure good and laterally uniform electrode contact, all self-potential monitoring locations consisted of a 0.30 m deep hole filled with a viscous bentonite slurry. The reference electrode was located in a pristine area of the site, and we followed standard practice in collecting and drift-correcting self-potential data with the roving electrode relative to this reference electrode [*Reynolds*, 1997; *Sharma*, 1997]. More specifically, all survey lines were connected with each other using numerous tie-in points, forming loops [*Naudet et al.*, 2004]. Loop-closure errors were re-distributed over the measurement stations in any particular loop, and were generally minute compared to the self-potential anomalies (several 100 millivolts) interpreted here. We adopted the methodology pioneered by *Naudet et al.* [2004] in isolating the streaming potential contribution to, and subtracting this contribution from, the total measured self-potential map.

Resistivity and induced polarization data are usually collected using dedicated instrumentation that switches automatically between series of quadripoles, where respectively two stainless-steel electrodes are used to inject the current at the ground surface and measure the ground's voltage response. Stainless-steel electrodes were readily available in the required quantity and expected to perform acceptably well [*LaBrecque and Daily*, 2008]; although we recognize that other electrode materials may have been preferable. Here we used an IRIS *Syscal R1Plus Switch 36* imaging system (www.iris-instruments.com) for data collection, with an array of 36 electrodes spaced 2 m apart, in the *Wenner* configuration, along the ten profiles spanned by the East-West nodes of the self-potential grid. The induced polarization and resistivity data were subsequently inverted tomographically in 2-D using the Res2Dinv software using the default inversion settings [*Loke and Barker*, 1996] inverting resistivity and IP concurrently, inferring the spatial distribution of resistivity and chargeability. Inversions typically converged within five iterations with a root mean square error of less than five percent. Chargeability expresses the magnitude of the induced polarization effect in the time domain, as manifested in a residual voltage after termination of current injection [*Slater and Lesmes*, 2002]. The induced polarization and resistivity data were interpolated in 3-D by the inverse distance method using the commercially available Voxler software (www.goldensoftware.com). The resistivity of in situ near surface materials is principally a function of the electrical properties of the fluids in the pore space. Chargeability is a function of both the pore fluid electrical properties and those of the interface between the solid matrix and the fluid-bearing pore space within it in the absence of continuous electronic conductors [*Slater and Lesmes*, 2002]. The important implication is that normalization of chargeability by resistivity can eliminate pore fluid effects and therefore emphasize the electrical properties of the solid matrix [*Slater and Lesmes*, 2002].

RESULTS

Reconnaissance Electromagnetic and Water and Soil Quality Surveys

The EM-31 and EM-38 bulk ground conductivity data, obtained prior to the intrusive investigation, identified a conductive anomaly of up to ~ 800 ms m^{-1} (Figure 1). Three trial pits (TPs 15, 32, 33; Figure 1) were excavated in this area to a maximum depth of 2.2 m. All three trial pits produced dark ashy metallic clinker and miscellaneous gravel sized particles with occasional fused iron to a depth of 1.7–1.8m where natural ground in the form of a clay aquiclude was encountered. This clay layer was found throughout the site and varied in thickness from 0.5m to 2.5m. The lithologies encountered were alluvial clays overlying interbedded sands and silts which lay on stiff glacial clays. Subsequent analysis of soils and perched waters from the fill material in TPs 15, 32 and 33 confirmed that contamination was not of concern in this area. Field measurements of the perched groundwater, at temporary installations using a multimeter and flow cell, revealed fluid electrical conductivities of 1210–1460 μS/cm conductivity, dissolved oxygen levels of 0.11–1.01 ppm, redox potentials (Eh) of 67–97mV, and pHs of 6.48–6.66. Significantly for our biogeophysical interpretation (section 5), it was concluded that the strong EM-31 anomaly most likely reflected a shallow aerobic perched water body that contains conductive materials as described above. The EM61 data (not presented) highlighted only discontinuous sections of pipework and reinforced concrete.

Contaminant Biogeochemistry Identified by the Intrusive Investigation

The contaminant of concern, or risk driver, for permeable reactive barrier implementation are polycyclic aromatic hydrocarbons which are effectively degraded in a biological permeable reactive barrier (Tables 2 and 3; the full range of organic contaminants is not listed for clarity). For simplicity, solvent extractable matter is presented as representative of the sum of organic contaminants of concern (Figure 3a). It is reported as the aliquot of the Soxhlet extraction of the soil sample, from which further organic fractions (mineral oil, non volatile aromatics) have been identified by chromatographic separation. Ammoniacal nitrogen, sulphate and total cyanide contaminants, commonly associated with the purification processes at former manufactured gas plants, are presented in aid of the biogeophysical interpretation. Groundwater field measurements (Eh and pH) from multilevel piezometers in boreholes are presented in Figure 2. Only a small proportion of indigenous microorganisms are likely to be culturable in a laboratory environment (about 0.1–10%), and replication of in situ conditions is almost impossible with bias playing a significant role in the cultivation of mixed microbial populations. Consequently an indication of the variety and composition of microbial species across the site was achieved using molecular genetic techniques (i.e., cloning and 16S rDNA gene sequencing). Variation in the microbial community structures was observed following the comparison of soil DNA samples and Table 4 summarizes the results of the cloning of 16SrDNA and sequencing experiments from DNA samples obtained from soils directly. Only unique sequences from over 100 clones are shown and this indicates that the Gram negative proteobacteria were found with β-proteobacteria (especially from the genus *Azoarcus)*, γ-proteobacteria (genus *Pseudomonas*), and δ-proteobacteria (genus *Geobacter*) being represented.

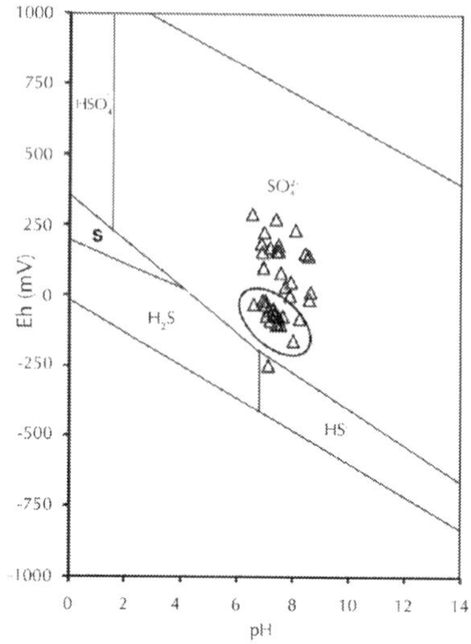

Figure 2: Eh versus pH scatterplot of groundwater samples from across the site. Samples from within the contaminant plume are enclosed in the ellipse.

	Solvent Extractable Matter	**Ammoniacal Nitrogen**	**Soluble Sulphate**	**Total Cyanide**	**Electrical Conductivity (μS/cm)**
Mean	9.5	120.5	806	5.6	1500
Median	4.0	20.3	661	0.3	1240
Mode	1.0	0.2[b]	2027.2	0.05[c]	
Standard Deviation	15.7	177.4	687	14.4	790
Minimum	1.0[d]	0.2[b]	52.4	0.05[c]	650
Maximum	83.0	726.4	2814.1	85.1	3290
Count	70	61	65	63	35
a All results in ppm except electrical conductivity μS/cm. Non detects replaced by method detection limit. b Ammoniacal nitrogen 0.2ppm. c Total cyanide 0.05ppm. d Solvent extractable matter 1ppm.					

Contaminant Transport Modeling

The modeled contaminant plume used a single source at depth from the area of the tar well. This was compared with groundwater monitoring from the intrusive phase of the site investigation. The spatial distribution of the groundwater data set indicates a source at depth (tar well) and various shallow sources originating from demolished structures at the west of site. Taking these shallow or surface sources from buildings on western site of the site into account there is still a good correlation between the modeled plume and measured solvent extractable matter in groundwater (Figure 3a). This provided assurance that the majority of groundwater contamination originates at depth (~5–7m below ground) from the tar well.

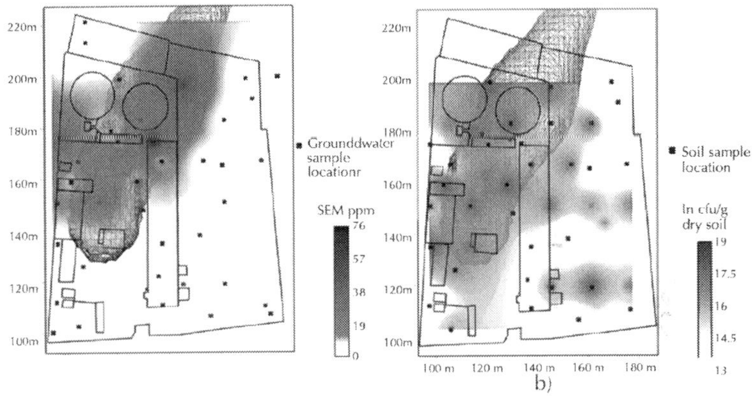

Figure 3: (a) Comparison of Solvent Extractable Matter (SEM) in groundwater as ppm (gray scale) with a modeled contaminant plume (black mesh) originating from the tar well at a depth of 5–7m. (b) Total viable microbial counts (ln cfu/g dry soil) as greyscale, note the microbial communities inferred from the total viable counts extend further to the east than the contaminant plume.

SP and IP Data

An exceptionally strong dipolar self potential anomaly (155 mV to +380 mV), with a sharp negative-to-positive polarity switch over an East-West distance of less than 5 m, dominated the northeastern section of the site (Figure 4). Following correction of our self-potential data for groundwater flow-induced streaming potentials (as explained by *Revil et al.* [2010]), the residual self-potential map reflects electrochemically and biologically generated self-potential signals. The spatial extent of the strong dipolar self-potential anomaly agrees approximately with that of the conductive EM-31 anomaly (Figure 4), and therefore with that of the inferred shallow aerobic perched water body containing conductive materials. Significantly for our interpretation (section 5), the sharp transition zone marking the self-potential polarity switch coincides spatially with the eastern margin of the active microbial communities as inferred from total viable counts from soil samples (Figure 3b).

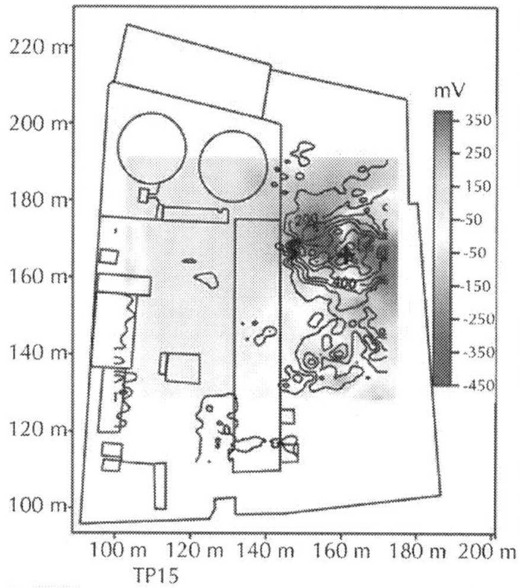

Figure 4: Dipolar self-potential anomaly (colors, mV) and contoured outlines of EM31 apparent conductivity (in mS/m). Note the sharp self-potential-polarity switch centered on Trial Pit 15 (labeled as a black cross).

Normalization of chargeabilities (section 3.4), by dividing our inverted induced polarization by the inverted resistivity data, allowed minimization of pore fluid electrical effects, including those of the inorganic conductive contaminants (ammonia and sulphate) in groundwater. We expect normalized chargeabilities to reflect buried metals, clays, bioprecipitation, mineralization or accumulations of microbes present in the pore space and attached to the solid matrix

[*Reynolds*, 1997; *Sharma*, 1997; *Slater and Lesmes*, 2002; *Abdel Aal et al.*, 2004; *Ntarlagiannis et al.*, 2005; *Davis et al.*, 2006; *Ntarlagiannis and Ferguson*, 2009]. Significantly for our interpretation (section 5), the spatial extent of the most prominent normalized-chargeability anomaly agrees well with that of the positive portion of the strong dipolar self-potential anomaly (Figure 5). Several more minor normalized-chargeability anomalies are observed in the western and southern sectors, and are attributed to metallic objects left on site during the installation of the permeable reactive barrier or other demolition works. The resistivity anomaly (Figure 5d) occurs at depth below the maximum depth of trial pit sampling, preventing a chemical or microbial benchmark. The anomaly may be related to a mobile non aqueous phase liquid moving at the base of the contaminant plume [*Sentenac et al.*, 2009]. Further intrusive work is warranted to benchmark this anomaly.

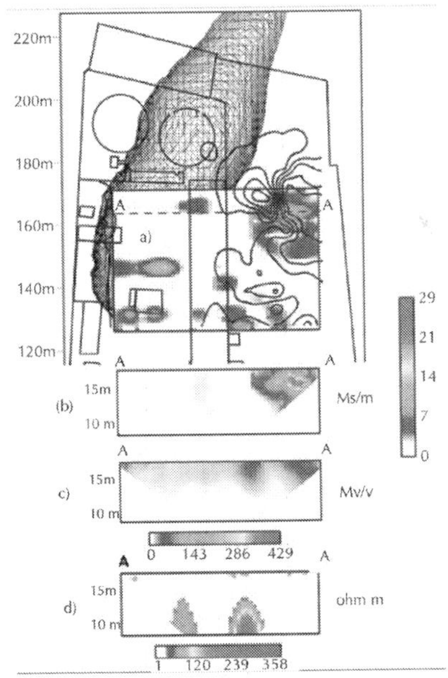

Figure 5: Plan of modeled contaminant plume (black mesh), contoured outlines of dipolar self-potential anomaly in mV (see Figure 4 for reference). (a) Chargeability normalized by resistivity (Ms/m) in color with 2D sections through line A-A . (b) Chargeability normalized by resistivity (Ms/m). (c) Chargeability (Mv/V). (d) Resistivity (Ohm m).

SYNTHESIS AND DISCUSSION

Conceptual Contaminant Biogeochemical Model

The tar well at the southwestern part of site is the dominant source of organic contaminants in groundwater. Groundwater flowing in a northerly direction [*Doherty et al.*, 2003] is contaminated at depth (5–7m) by residual dense non aqueous phase liquids from the tar well. This plume is contained at depth by a thin clay aquiclude (0.5–2.5m) that covers the site. Ammonia was present as a liquid gasworks waste, known as *ammoniacal liquor* [*Hatheway*, 2002], and was usually disposed of in the tar well. Cyanide and sulphate compounds present as solid wastes were often reused as fill materials around the site [*Desrochers*, 2009]. In the conceptual contaminant transport model of the site, ammonia occupies the same source area as the organic contaminants at the tar well. Cyanide and sulphate source areas are closer to the surface and are identified with structures or waste fill material. Elevated contaminant levels of ammonium of up to 726 ppm across the site, together with Eh pH measurements (Figure 2), suggest that the groundwater contaminants are in an anaerobic environment. Ammonium undergoes nitrification in an aerobic environment to form nitrate. *Broholm et al.* [1998] and *Torstensson et al.* [1998] noted that microbial oxidation of the ammonium in a former manufactured gas plant-type plume delineated the aerobic-anaerobic boundary within the contaminated groundwater. The elevated ammonium concentrations suggest that any available dissolved oxygen within the plume would be quickly depleted when nitrification of ammonium occurs. The molecular genetic techniques used to analyze soil bacteria at the site indicated that microorganisms of the proteobacteria were most common; especially those known to be associated with denitrification (Table 4). This indicates that the anaerobic utilization of nitrate was most likely. Ammonia dominates the contaminant biogeochemistry, preferentially sequestering dissolved oxygen. This in turn prevents aerobic biodegradation of organic compounds. Additional electron acceptors (sulphate reduction and methanogenesis) are also suppressed, as suggested by the groundwater Eh versus pH values (Figure 2) until the nitrogen electron acceptors (ammonia, nitrate and nitrite) are exhausted. Compound Specific Isotope Analysis work using Gas Chromatography - Mass Selective Detection - Isotope Ratio Mass Spectrometry from trial pit samples found that there was significant $\delta^{13}C$ variation depending on the matrix the contaminant was sampled from, and overall there was no conclusive evidence for widespread bioattenuation [*Hall*, 1999]. However, *Hall* [1999] noted $\delta^{13}C$ fractionation of residual aliphatic and phenolic compounds in groundwater. The majority of aerobic degradation of aromatic compounds is effectively stalled through oxygen depletion and possibly by the presence of ammonium, although there can still be a degree of anaerobic degradation of aliphatic and phenolic compounds. Total viable counts of aerobic microorganisms from soil samples [*Ferguson et al.*, 2003] reproduced here correlates well with contamination in groundwater (Figures 3a and 3b). Localized areas of low microbial viable counts at the plume can be attributed to surface structures, such as the purifiers and waste spoils that were localized sources of cyanide contamination and high pH. Relatively low microbial viable counts were also recorded in the pristine southeastern corner of the site (local grid 110N 170W) where no contaminating activities occurred, and where the SP reference electrode was located. A smaller separate set of viable counts (local grid 115N 160W) not based around the plume and tar well has been attributed to minor surface spills. This soil total viable count data is presented as inference of an active microbial community surrounding the organic contaminant plume [*Bakermans and Madsen*, 2000]. Recent work on similar sites suggests that microbial diversity is at a maximum at the plume edge, whereas only specific degraders can excel in the center of contaminant plumes where toxicity is greatest [*Ferguson et al.*, 2007]. The microbial community around the contaminant plume can be conceptualized as utilizing the contaminant plume as an energy source, mediating redox reactions and catalyzing contaminant degradation to provide a source of electrons (Figures 3a and 3b).

Microbial Fuel Cell as an Alternative to a Biogeobattery

Field measurements revealed redox values ranging from +97 mV in the body of aerobic perched water to −161 mV in the anaerobic groundwater contaminant plume. The redox range between these bodies is narrow compared to the dipolar SP anomaly (>800 mV peak-to-peak). This narrow range of redox potential is not a large enough source mechanism to drive the observed electrical current flow as we would expect in a geochemically dominated, redox driven, geobattery model [*Arora et al.*, 2007] or biogeobattery model [*Revil et al.*, 2010], or to promote the precipitation of metal sulphides that contribute to IP responses reported by *Williams et al.* [2009] (Figure 2). The field SP response is an order of magnitude greater than the observed response from the corrosion of metallic objects as noted in laboratory experiments by *Castermant et al.* [2008], suggesting it is unlikely that the SP anomaly is due to corrosion of metallic debris in the perched water alone. In the absence of an alternative plausible explanation, we believe that the strong dipolar self-potential anomaly (Figure 4) reflects, instead, a microbial fuel cell [*He and Angenent*, 2006], which is theoretically well-founded [*Revil et al.*, 2010]. Microbial fuel cells generate electrical current by utilizing microbes to catalyze organic material producing electrons [*Du et al.*, 2007]; in this case the microbes catalyze the contaminant plume. As explained below, this cell is inferred to be characterized by a biological anode associated with the active microbial soil communities around the contaminant plume (as reflected by molecular genetic techniques; Table 4 and total viable counts; Figure 3b), an abiotic cathode associated with the metallic infill congregated within the aerobic body of perched water (as reflected by the pronounced normalized-chargeability anomaly; Figure 5), and a locally punctured clay aquiclude around trial pit 15 that facilitates transfer of charge and acts as an electronic conductor interconnecting the biological anode and the abiotic cathode. We use the term microbial fuel cell rather than biogeobattery due to the fact that our conceptual model proposes a biological anode and abiotic cathode rather than a series of biologically mediated electron acceptors acting as the anode and cathode.

Biological Anode

The anaerobic conditions in the organic contaminant plume can readily provide negatively charged conditions (Figure 2). The active anaerobic microbial soil communities could furnish an efficient mechanism allowing electron flow in the areas adjacent to the contaminant plume (Figure 3b). There is still much debate over mechanisms that allow electron transfer from microbial cells [*Lovley*, 2008a, 2008b]. The mechanisms can be generally split in two categories, transfer of electrons using nanowires or pili [*Reguera et al.*, 2005; *Gorby et al.*, 2006;*Ntarlagiannis et al.*, 2007; *Gorby et al.*, 2008], or transfer of electrons using soluble recalcitrant organic electron shuttles such as flavin [*Velasquez-Orta et al.*, 2009] or naturally occurring humic materials [*Newman and Kolter*, 2000]. Site specific microbial data (cloning and 16S rDNA gene sequencing) suggest humic reducing microbes are established in soil samples (Table 4). In all of the above, the transfer of scale from mechanisms occurring at the cell to measurements in the field is important. The models proposed by *Revil et al.* [2010] allow the aggregation of microscale electron transfer (either between cells or between cells and precipitates/aquifer media) to create macroscopic dipoles. Specifically, this would allow the microbial soil communities on the eastern side of the contaminant plume (Figure 3b) to act as an anode. This area of the microbial soil communities are marked by negative SP values up to the sharp polarity switch at the plume's eastern margin. The site-specific microbial data (cloning and 16S rDNA gene sequencing) reveal that electrons cannot only come from microbial utilization of contaminant organic matter [*Atekwana et al.*, 2005; *Du et al.*, 2007; *Revil et al.*, 2010], but also from mechanisms that involve ammonia-oxidizing chemoautotrophs such as β-Proteobacetria (*Nitrosolobus spp*) [*Head et al.* 1993]

$$NH^{4+} + O_2 \rightarrow NO_2^- + 4H^+ + 2e- \quad (1)$$

Many nitrifying or nitrate reducing bacteria were established as common in soil at the site (Table 4) and this suggests nitrate production and nitrification is also very important:

$$2NO^{2-} + H_2O \rightarrow 2NO^{3-} + 2H^+ + 2e- \quad (2)$$

Equations (1) and (2) are consistent with the strong possibility of an effective biologically dominated anode that couples the contaminant plume with the surrounding microbial architecture, including ammonia oxidizers, nitrate reducers and organic contaminant degraders acting as the *electrode reducers* [*Lovley*, 2008a].

Abiotic Cathode

In the search for an abiotic cathode we recall that the spatial extent of most the prominent normalized-chargeability anomaly agrees well with the positive portion of the strong dipolar self-potential anomaly (Figure 5). In the absence of an alternative plausible explanation, we postulate that the mechanism generating this anomaly could also act as the abiotic cathode of our microbial fuel cell. Indeed, the normalized-chargeability anomaly probably delineates the ash, clinker and iron compounds that were discovered and back-filled during trial-pitting. These compounds are congregated within the shallow body of perched groundwater delineated by our EM-31 data (compareFigures 1 and 5). We therefore believe that geochemical reactions within these oxygen-rich waters are oxidizing the waste-iron compounds, effectively acting as the abiotic cathode of our microbial fuel cell. We do note that microbiological analysis has also identified denitrifying bacteria which could potentially compete with the abiotic cathode for electrons.

Interconnecting Electronic Conductor

Trial-pitting (TPs 15 and 32) confirmed that a shallow body of perched aerobic water and backfilled conductive waste overlying a layer of clay, as delineated by the EM-31 data (Figures 1 and 4). We interpret this as the abiotic cathode delineated by the normalized-chargeability anomaly that is also present (Figure 5). The clay layer separating the perched water and contaminant plume effectively divides these bodies into an abiotic cathode chamber (perched water) and a biological anode chamber (contaminant plume). The clay layer was thinned and probably punctured during the excavation of TP 15 which sits in the center of the SP dipole (Figures 4 and 5), connecting the oxidizing body of perched waters above it with the reducing environment beneath it. A microbial fuel cell requires a cation-exchange membrane that allows cations or protons to diffuse and thus allow the anode to operate efficiently [*Rozendal et al.*, 2006]. This punctured clay layer, backfilled with conductive waste, may act as an exchange membrane or electronic conductor connecting the contaminant plume, the biological anode of our microbial fuel cell, to the normalized-chargeability anomaly representing the abiotic cathode within the shallow body of perched water (Figures 1 and 5). This would support a situation where the contaminant plume and electrode reducers provide electrons that reduce the waste-iron compounds to Fe (II):

$$Fe^{3+} + e^{-} \rightarrow Fe^{2+} \tag{3}$$

The aerobic water above could oxidize the reduced iron to Fe (III):

$$4Fe^{2+} + O_2 + 4H^{+} \rightarrow 4Fe^{3+} + 2H_2O \tag{4}$$

The waste-iron compounds backfilled in TPs 15, 32 and 33 (Figure 1) would, thus, act as electrode mediators between oxygen, derived from the aerobic perched waters (dissolved oxygen measured in TP15 ranged from 0.11 to 1.01 ppm and redox potential was 67–97 mV, section 4.1), and the abiotic cathode, where oxygen is the terminal electron acceptor [*Park and Zeikus*, 2003; *He and Angenent*, 2006]:

$$14e^{-} + 3.5O_2 + 14H^{+} \rightarrow 7H_2O \tag{5}$$

Implications for the Sustainable Remediation of Contaminated Land

The use of multiple geophysics methods (electromagnetic, self-potential, induced polarization and resistivity) alongside chemical and microbiological analysis has been invaluable in developing and quantifying site conceptual models and the management of a site remediation strategy. The use of electrical geophysical methods as remediation monitoring tools has provided additional insight to a complex biogeochemical environment. The large self-potential response is probably generated by contaminant biodegradation providing a source of electrons to the fuel cell. This has implications for our ability to actively monitor biodegradation in contaminant plumes. Demonstration of ongoing biodegradation is a requirement for remediation methods such as monitored natural attenuation. This is often the most technically difficult, expensive and intrusive aspect of the monitoring process. The electrical geophysical methods show promise as inexpensive, non intrusive, real time methods. These are the qualities of 'sustainable remediation' tools [*Spira et al.*, 2006; *U.S. Sustainable Remediation Forum*, 2009] that are required if contaminated land issues are to be managed efficiently in the future. Further work is still required to develop field scale abiotic self-potential cathodes to complete the flow of electrons from biodegrading plumes that act as bio-anodes or to enhance biodegradation [*Zhang et al.*, 2010]. The use of electrical geophysical methods could have further applications in more controlled engineered environments where biodegradation or microbial activity occurs, for example, municipal wastewater systems, landfill and leachate treatment, anaerobic digestion of wastes using mechanical - biological treatment and monitoring of industrial biofouling.

CONCLUSIONS

The revised conceptual model of the site considers a contaminant plume from the tar well. The plume consists predominately of organic contaminants and inorganic ammonia disposed of below the water table. Total viable counts from soil samples, as well as cloning and 16S rDNA sequencing, provides assurance that the plume has a diverse microbial community around it. An exceptionally strong, dipolar self-potential anomaly (>800 mV peak-to-peak after correction for streaming potential) is inferred to be generated by a microbial fuel cell operating in the eastern sector of the site. The contaminant biogeochemistry is dominated by ammonia, oxidation to nitrate and subsequent reduction along with some biodegradation of aliphatic and phenolic compounds. This reducing environment and associated biodegradation around the contaminant plume is proposed as an anode of the microbial fuel cell producing electrons. A congregation of waste-iron compounds, back-filled after trial-pitting and situated within a shallow body of aerobic perched water, is inferred to act as the cell's abiotic cathode. The perched-water body and the waste-iron compounds within it were respectively delineated as strong EM-31 and normalized-chargeability anomalies. This body of aerobic perched water, and thus the waste-iron compounds, overlie a clay layer which has been thinned and probably punctured during the intrusive site investigation. This thinned and punctured clay could now act as a permeable membrane allowing ion and electron transport between the electron-providing anode and the oxidizing cathode, thus acting as an interconnecting electronic conductor of the microbial fuel cell. The transport of electrons from the anode is probably provided by an aggregation of microscale cell to cell, or cell to aquifer electron transfer to the punctured aquiclude; where geochemical reduction and oxidation of iron compounds takes place. From our field data we cannot attribute a specific mechanism of electron transport from microbial communities; possibilities are electron shuttling from naturally occurring humic compounds, extracellular transfer across pili or a combination of the two. Further work is warranted to define the modes of electron transport and/or electronic conductors. In this case, geophysics provided assurance that microbially dominated conditions around the plume were present and active. The spatial distribution of the geophysical measurements around the contaminant plume and perched water suggests that microbial fuel cells do function without assistance in anthropogenic environments outside of the laboratory. This has future applications in terms of design, engineering and monitoring of biological systems as proposed by *Curtis et al.* [2003].

ACKNOWLEDGMENTS

This research was supported by EPSRC grant GR/M89768/01, the Department for the Environment Northern Ireland, the Questor Centre, and Keller Ground Engineering Ltd. A. Brown and A. de Jong collected the EM data. David Leemon and John Patterson collected the SP, IP and resistivity data, and Valerie Irvine supported the microbiological analyses. We would also like to thank the constructive and helpful comments of the reviewers.

REFERENCES

1. Abdel Aal, G. Z., E. A. Atekwana, L. D. Slater, and E. A. Atekwana (2004), Effects of microbial processes on electrolytic and interfacial electrical properties of unconsolidated sediments, *Geophys. Res. Lett.*, 31, L12505, doi:10.1029/2004GL020030.
2. Arora, T., N. Linde, A. Revil, and J. Castermant (2007), Non-

intrusive characterization of the redox potential of landfill leachate plumes from self-potential data, *J. Contam. Hydrol.*, 92, 274–292, doi:10.1016/j.jconhyd.2007.01.018.

3. Atekwana, E. A., E. Atekwana, F. D. Legall, and R. V. Krishnamurthy (2005), Biodegradation and mineral weathering controls on bulk electrical conductivity in a shallow hydrocarbon contaminated aquifer, *J. Contam. Hydrol.*, 80, 149–167, doi:10.1016/j.jconhyd.2005.06.009.

4. Bakermans, C., and E. L. Madsen (2000), Use of substrate responsive-direct viable counts to visualize naphthalene degrading bacteria in a coal tar-contaminated groundwater microbial community, *J. Microbiol. Methods*, 43(2), 81–90, doi:10.1016/S0167-7012(00)00210-4.

5. Bigalke, J., and E. W. Grabner (1997), The Geobattery model: A contribution to large scale electrochemistry, *Electrochim. Acta*, 42(23–24), 3443–3452, doi:10.1016/S0013-4686(97)00053-4.

6. Broholm, M. M., I. Jones, D. Torstensson, and E. Arvin (1998), Groundwater contamination from a coal carbonisation plant, in*Contaminated Land and Groundwater: Future Directions*, edited by D. N. Lerner, and N. Walton, pp. 159–165, Geol. Soc., London.

7. Castermant, J., C. A. Mendonça, A. Revil, F. Trolard, G. Bourrié, and N. Linde (2008), Redox potential distribution inferred from self-potential measurements associated with the corrosion of a burden metallic body, *Geophys. Prospect.*, 56(2), 269–282, doi:10.1111/j.1365-2478.2007.00675.x.

8. Curtis, T. P., I. M. Head, and D. W. Graham (2003), Theoretical ecology for engineering biology, *Environ. Sci. Technol.*, 37(3),64A–70A, doi:10.1021/es0323493.

9. Davis, C. A., E. A. Atekwana, E. Atekwana, L. D. Slater, S. Rossbach, and M. R. Mormille (2006), Microbial growth and biofilm formation in geologic media is detected with complex conductivity measurements, *Geophys. Res. Lett.*, 33, L18403, doi:10.1029/2006GL027312.

10. Desrochers, P. (2009), Does the invisible hand have a green thumb? Incentives, linkages, and the creation of wealth out of industrial waste in Victorian England, *Geogr. J.*, 175(1), 3–16, doi:10.1111/j.1475-4959.2008.00315.x.

11. Doherty, R., U. S. Ofterdinger, Y. Yang, K. Dickson, and R. M. Kalin (2003), Observed and modelled hydraulic aquifer response to slurry wall installation at the former Gasworks Site, in *Advanced Groundwater Remediation: Active and Passive Technologies*, edited by F. G. Simon, T. Meggyes, and C. M. McDonald, pp. 239–249, Thomas Telford, Portadown, U. K.,
12. Du, Z., H. Li, and T. Gu (2007), A state of the art review on microbial fuel cells: A promising technology for wastewater treatment and bioenergy, *Biotechnol. Adv.*, 25(5), 464–482, doi:10.1016/j.biotechadv.2007.05.004.
13. Ferguson, A. S., R. Doherty, M. J. Larkin, R. M. Kalin, V. Irvine, and U. S. Ofterdinger (2003), Toxicity assessment of a former manufactured gas plant, *Bull. Environ. Contam. Toxicol.*, 71(1), 21–30, doi:10.1007/s00128-003-0125-y.
14. Ferguson, A. S., et al. (2007), Microbial analysis of soil and groundwater from a gasworks site and comparison with a sequenced biological reactive barrier remediation process, *J. Appl. Microbiol.*, 102(5), 1227–1238, doi:10.1111/j.1365-2672.2007.03398.x.
15. Fournier, C. (1989), Spontaneous potentials and resistivity surveys applied to hydrogeology in a volcanic area: Case history of the Chaine des Puys (Puy-de-Dome, France), *Geophys. Prospect.*, 37, 647–668, doi:10.1111/j.1365-2478.1989.tb02228.x.
16. Gibert, O., A. S. Ferguson, R. M. Kalin, R. Doherty, K. W. Dickson, K. L. McGeough, J. Robinson, and R. Thomas (2007), Performance of a sequential reactive barrier for bioremediation of coal tar contaminated groundwater, *Environ. Sci. Technol.*, 41(19),6795–6801, doi:10.1021/es071527f.
17. Gorby, Y. A., et al. (2006), Electrically conductive bacterial nanowires produced by Shewanella oneidensis strain MR-1 and other microorganisms, *Proc. Natl. Acad. Sci. U. S. A.*, 103(30), 11,358–11,363, doi:10.1073/pnas.0604517103.
18. Gorby, Y., J. Mclean, A. Korenevsky, K. Rosso, M. Y. El-Naggar, and T. J. Beveridge (2008), Redox-reactive membrane vesicles produced by *Shewanella*, *Geobiology*, 6(3), 232–241, doi:10.1111/j.1472-4669.2008.00158.x.

19. Hall, J. (1999), Stable isotope methods for monitoring bioremediation of contaminated land, Ph.D. thesis, Queen's Univ. of Belfast,Belfast.

20. Hatheway, A. W. (2002), Geoenvironmental protocol for site and waste characterization of former manufactured gas plants: Worldwide remediation challenge in semi-volatile organic wastes, *Eng. Geol.*, 64(4), 317–338, doi:10.1016/S0013-7952(01)00097-7.

21. He, Z., and L. T. Angenent (2006), Application of bacterial biocathodes in microbial fuel cells, *Electroanalysis*, 18(19–20), 2009–2015, doi:10.1002/elan.200603628.

22. Head, I. M., W. D. Hiorns, M. T. Embley, A. J. McCarthy, and J. R. Saunders (1993), The phylogeny of autotrophic ammonia-oxidizing bacteria as determined by analysis of 16S ribosomal RNA gene sequences, *J. Gen. Microbiol.*, 139, 1147–1153.

23. Kalin, R. M. (2004), Engineered passive bioreactive barriers: Risk-managing the legacy of industrial soil and groundwater pollution,*Curr. Opin. Microbiol.*, 7, 227–238, doi:10.1016/j.mib.2004.04.014.

24. Kulessa, B., B. Hubbard, and G. H. Brown (2003), Cross-coupled flow modeling of coincident streaming and electrochemical potentials and application to subglacial self-potential data, *J. Geophys. Res.*, 108(B8), 2381, doi:10.1029/2001JB001167.

25. LaBrecque, D., and W. Daily (2008), Assessment of measurement errors for galvanic-resistivity electrodes of different composition,*Geophysics*, 73, F55–F64, doi:10.1190/1.2823457.

26. Leu, J. Y., C. P. McGovern-Traa, A. J. R. Poter, W. J. Harris, and W. A. Hamilton (1998), Identification and phylogenetic analysis of thermophilic sulphate-reducing bacteria in oil fuel samples by 16S rDNA gene cloning and sequencing, *Anaerobe*, 4, 165–174, doi:10.1006/anae.1998.0156.

27. Linde, N., A. Revil, A. Bolève, C. Dagès, J. Castermant, B. Suski, and M. Voltz (2007), Estimation of the water table throughout a catchment using self-potential and piezometric data in a Bayesian framework, *J. Hydrol.*, 334(1–2), 88–98, doi:10.1016/j.jhydrol.2006.09.027.

28. Loke, M. H., and R. D. Barker (1996), Rapid least-squares inversion of apparent resistivity pseudosections by a quasi-Newton method,*Geophys. Prospect.*, 44(1), 131–152, doi:10.1111/j.1365-2478.1996.tb00142.x.

29. Lovley, D. R. (2008a), The microbe electric: Conversion of organic matter to electricity, *Curr. Opin. Biotechnol.*, 19(6), 564–571, doi:10.1016/j.copbio.2008.10.005.

30. Lovley, D. R. (2008b), Extracellular electron transfer: Wires, capacitors, iron lungs, and more, *Geobiology*, 6(3), 225–231, doi:10.1111/j.1472-4669.2008.00148.x.

31. Maineult, A., Y. Bernabé, and P. Ackerer (2006), Detection of advected, reacting redox fronts from self-potential measurements, *J. Contam. Hydrol.*, 86(1–2), 32–52, doi:10.1016/j.jconhyd.2006.02.007.

32. Minsley, B. J., J. Sogade, and F. D. Morgan (2007), Three-dimensional self-potential inversion for subsurface DNAPL contaminant detection at the Savannah River Site, South Carolina, *Water Resour. Res.*, 43, W04429, doi:10.1029/2005WR003996.

33. Naudet, V., and A. Revil (2005), A sandbox experiment to investigate bacteria-mediated redox processes on self-potential signals,*Geophys. Res. Lett.*, 32, L11405, doi:10.1029/2005GL022735.

34. Naudet, V., A. Revil, E. Rizzo, J. Y. Bottero, and P. Bagassat (2004), Groundwater redox conditions and conductivity in a contaminant plume from geoelectrical investigations, *Hydrol. Earth Syst. Sci.*, 8(1), 8–22, doi:10.5194/hess-8-8-2004.

35. Newman, D. K., and R. Kolter (2000), A role for excreted quinones in extracellular electron transfer, *Nature*, 405(6782), 94–97, doi:10.1038/35011098.

36. Ntarlagiannis, D., and A. Ferguson (2009), SIP response of artificial biofilms, *Geophysics*, 74(1), A1–A5, doi:10.1190/1.3031514.

37. Ntarlagiannis, D., N. Yee, and L. D. Slater (2005), On the low-frequency electrical polarization of bacterial cells in sands, *Geophys. Res. Lett.*, 32, L24402, doi:10.1029/2005GL024751.

38. Ntarlagiannis, D., E. A. Atekwana, E. A. Hill, and Y. Gorby (2007), Microbial nanowires: Is the subsurface "hardwired"? *Geophys. Res. Lett.*, 34, L17305, doi:10.1029/2007GL030426.

39. Park, D. H., and J. G. Zeikus (2003), Improved fuel cell and electrode designs for producing electricity from microbial degradation,*Biotechnol. Bioeng.*, 81(3), 348–355, doi:10.1002/bit.10501.

40. Personna, Y. R., D. Ntarlagiannis, L. Slater, N. Yee, M. O'Brien, and S. Hubbard (2008), Spectral induced polarization and electrodic potential monitoring of microbially-mediated iron sulfide transformations, *J. Geophys. Res.*, 113, G02020, doi:10.1029/2007JG000614.

41. Petiau, G. (2000), Second generation of lead-lead chloride electrodes for geophysical applications, *Pure Appl. Geophys.*, 157,357–382, doi:10.1007/s000240050004.

42. Reguera, G., K. D. McCarthy, T. Mehta, J. S. Nicoll, M. T. Tuominen, and D. R. Lovley (2005), Extracellular electron transfer via microbial nanowires, *Nature*, 435(7045), 1098–1101, doi:10.1038/nature03661.

43. Revil, A., V. Naudet, J. Nouzaret, and M. Pessel (2003), Principles of electrography applied to self-potential electrokinetic sources and hydrogeological applications, *Water Resour. Res.*, 39(5), 1114, doi:10.1029/2001WR000916.

44. Revil, A., F. Trolard, G. Bourrié, J. Castermant, A. Jardani, and C. A. Mendonça (2009), Ionic contribution to the self-potential signals associated with a redox front, *J. Contam. Hydrol.*, 109(1–4), 27–39, doi:10.1016/j.jconhyd.2009.07.008.

45. Revil, A., C. A. Mendonça, E. A. Atekwana, B. Kulessa, S. Hubbard, and K. Bohlen (2010), Understanding biogeobatteries: Where geophysics meets microbiology, *J. Geophys. Res.*, 115, G00G02, doi:10.1029/2009JG001065.

46. Reynolds, J. M. (1997), *An Introduction to Applied and Environmental Geophysics*, John Wiley, Chichester, U. K.,

47. Rozendal, R. A., H. V. M. Hamelers, and C. J. N. Buisman (2006), Effects of membrane cation transport on pH and microbial fuel cell performance, *Environ. Sci. Technol.*, 40(17), 5206–5211, doi:10.1021/es060387r.

48. Sambrook, J., T. Maniatis, and E. F. Fritsch (1989), *Molecular Cloning: A Laboratory Manual*, 2nd ed., Cold Spring Harbor Lab., New York.,

49. Sandberg, S. K., L. D. Slater, and R. Versteeg (2002), An integrated geophysical investigation of the hydrogeology of an anisotropic unconfined aquifer, *J. Hydrol.*, 267(3–4), 227–243, doi:10.1016/S0022-1694(02)00153-1.

50. Sato, M., and H. M. Mooney (1960), The electrochemical mechanism of sulfide self-potentials, *Geophysics*, 25(1), 226–249, doi:10.1190/1.1438689.

51. Sentenac, P., A. Montinaro, and B. Kulessa (2009) Diesel transport monitoring in simulated unconfined aquifers using miniature resistivity arrays, *Environ. Earth Sci.*, doi:10.1007/s12665-009-0325-9, in press.

52. Sharma, P. (1997), *Environmental and Engineering Geophysics*, Cambridge Univ. Press, Cambridge, U. K.,

53. Slater, L., and A. Binley (2006), Synthetic and field-based electrical imaging of a zerovalent iron barrier: Implications for monitoring long-term barrier performance, *Geophysics*, 71(5), B129–B137, doi:10.1190/1.2235931.

54. Slater, L. D., and D. Lesmes (2002), IP interpretation in environmental investigations, *Geophysics*, 67, 77–88, doi:10.1190/1.1451353.

55. Slater, L., D. Ntarlagiannis, Y. R. Personna, and S. Hubbard (2007), Pore-scale spectral induced polarization signatures associated with FeS biomineral transformations, *Geophys. Res. Lett.*, 34, L21404, doi:10.1029/2007GL031840.

56. Spira, Y., J. Henstock, P. Nathanail, D. Müller, and D. Edwards (2006), A European approach to increase innovative soil and groundwater remediation technology applications, *Rem. J.*, 16(4), 81–96, doi:10.1002/rem.20103.

57. Torstensson, D., S. F. Thornton, M. M. Broholm, and D. N. Lerner (1998), Hydrochemistry of pollutant attenuation in groundwater contaminated by coal tar wastes, in *Contaminated Land and Groundwater: Future Directions*, edited by D. N. Lerner, and N. Walton, pp. 149–157, Geol. Soc., London.

58. U.S. Sustainable Remediation Forum (2009), Sustainable remediation white paper—Integrating sustainable principles, practices, and metrics into remediation projects, *Rem. J.*, 19(3), 5–114, doi:10.1002/rem.20210.

59. Velasquez-Orta, S., I. Head, T. Curtis, K. Scott, J. Lloyd, and H. von Canstein (2009), The effect of flavin electron shuttles in microbial fuel cells current production, *Appl. Microbiol. Biotechnol.*, 85, 1373–1381, doi:10.1007/s00253-009-2172-8.

60. Williams, K. H., A. Kemna, M. J. Wilkins, J. Druhan, E. Arntzen, A. L. N'Guessan, P. E. Long, S. S. Hubbard, and J. F. Banfield (2009),Geophysical monitoring of coupled microbial and geochemical processes during stimulated subsurface bioremediation, *Environ. Sci. Technol.*, 43(17), 6717–6723, doi:10.1021/es900855j.

61. Zhang, T., S. M. Gannon, K. P. Nevin, A. E. Franks, and D. R. Lovley (2010), Stimulating the anaerobic degradation of aromatic hydrocarbons in contaminated sediments by providing an electrode as the electron acceptor, *Environ. Microbiol.*, 12(4),1011–1020, doi:10.1111/j.1462-2920.2009.02145.x.

Chapter 8

Soil Engineering in Vivo: Harnessing Natural Biogeochemical Systems for Sustainable, Multi-functional Engineering Solutions

Jason T. DeJong,[1,2,*] Kenichi Soga,[3] Steven A. Banwart,[4] W. Richard Whalley,[5] Timothy R. Ginn,[1] Douglas C. Nelson,[1] Brina M. Mortensen,[1] Brian C. Martinez,[1] and Tammer Barkouki[1]

[1]Department of Civil and Environmental Engineering, University of California, One Shields Avenue, Davis, CA 95616, USA

[2]Institute for Advanced Studies, The University of Western Australia, 35 Stirling Highway Crawley, Perth Western Australia 6009, Australia

[3]Department of Engineering, University of Cambridge, Cambridge, UK

[4]Kroto Research Institute, University of Sheffield, Sheffield, UK

[5]Rothamsted Research, Harpenden, West Common, Hertfordshire AL5 2JQ, UK

ABSTRACT

Carbon sequestration, infrastructure rehabilitation, brownfields clean-up, hazardous waste disposal, water resources protection and global warming—these twenty-first century challenges can neither be solved by the high-energy consumptive practices that hallmark industry today, nor by minor tweaking or optimization of these processes. A more radical, holistic approach is required to develop the sustainable solutions society needs. Most of the above challenges occur within, are supported on, are enabled by or grown from soil. Soil, contrary to conventional civil engineering thought, is a living system host to multiple simultaneous processes. It is proposed herein that 'soil engineering *in vivo*', wherein the natural capacity of soil as a living ecosystem is used to provide multiple solutions simultaneously, may provide new, innovative, sustainable solutions to some of these great challenges of the twenty-first century. This requires a multi-disciplinary perspective that embraces the science of biology, chemistry and physics and applies this knowledge to provide multi-functional civil and environmental engineering designs for the soil environment. For example, can native soil bacterial species moderate the carbonate cycle in soils to simultaneously solidify liquefiable soil, immobilize reactive heavy metals and sequester carbon—effectively providing civil engineering functionality while clarifying the ground water and removing carbon from the atmosphere? Exploration of these ideas has begun in earnest in recent years. This paper explores the potential, challenges and opportunities of this new field, and highlights one biogeochemical function of soil that has shown promise and is developing rapidly as a new technology. The example is used to propose a generalized approach in which the potential of this new field can be fully realized.

INTRODUCTION

Carbon sequestration, infrastructure rehabilitation, brownfields clean-up, hazardous waste disposal, water resources protection and global warming—these twenty-first century challenges, which exist worldwide and impact all, are too large, complicated and interwoven to be solved with the high-energy consumptive practices that hallmark industry today. Engineering by brute force at the scale required is not possible, the totality of resources not available and their implementation unaffordable [1,2].

Most of the above challenges occur within, are supported on, are enabled by or grown from soil. Conventional perspective views soil as an infinite resource composed of separable, in fact isolated, functions. Geotechnical engineering considers the strength and deformation of soil; environmental engineering considers contaminants within the soil; plant science considers suitability for plant growth and to maximize the productivity of crops. Soil is a finite resource and it commands our stewardship [3,4].

The gravity of the situation is sinking in. Sustainability concepts and soil reuse planning are being implemented for select practices in first world countries. New construction is evaluated against carbon footprint, energy efficiency and total life-cycle analysis criteria (e.g. [5,6]). Unfortunately, the burgeoning world population demands still outpace the 'savings' gained by these reformations, such that the pace of consumption (and associated environmental impact) continues to increase. A more radical, holistic approach is required.

Herein, the potential for 'soil engineering *in vivo*', for soil to be considered as a living, active system that has the capacity to provide multiple sustainable solutions simultaneously, is explored. This requires a multi-disciplinary perspective that embraces the science of biology, chemistry and physics, and applies this knowledge to multi-functional civil and environmental engineering applications. For example, can native soil bacterial species moderate the carbonate cycle in soils to simultaneously solidify liquefiable soil, immobilize reactive heavy metals and sequester carbon—effectively providing civil engineering functionality while clarifying the ground water and removing carbon from the atmosphere? Exploration of these ideas has begun in earnest in recent years, and this paper explores the potential, challenges and opportunities of this new field, highlights one biogeochemical function of soil that has shown promise and is developing rapidly, and proposes a generalized approach in which the potential of this new field can be fully realized.

ENVISIONING THE POSSIBILITIES

Visions of the potential are emerging more rapidly than technical capabilities. Green Wall Sahara, an anti-desertification initiative that consists of stabilizing a 6000 km reach of sand dunes across northern Africa has been proposed by Magnus Larsson ([7]; figure 1*a*). One-quarter of the world's land mass, which is home to more than one billion people across 100 countries is threatened by desertification [8]. Specifically, Mauritania's capitol, Nouakchott, is on the verge of succumbing to this threat (figure 1*b*), and in Nigeria the desert is moving southward 600 m yr^{-1}. Larssons's idea is impossible using conventional soil-solidification measures based on concrete and chemicals, but stimulating and harnessing a natural process inherent to living soil—such as the bio-mediated calcifiction process discussed herein—might realize this dream.

Figure 1: (*a*) Proposed Great Wall Sahara treatment [7] and (*b*) desert encroachment towards Nouakchott, Mauritania.

Biologically aided carbon sequestration enabled by pedogenic carbonate precipitation, which occurs near plant root exudates in highly alkaline soil, may be an untapped naturally occurring reservoir primed to address this global issue. Plants exude 10–30% of the carbon captured from the atmosphere, by photosynthesis, through their roots and associated mycorrhizal fungal associations ([9,10] and included references). Any compound in the root tissue may be released into the soil in the form of an exudate (see [11]); thus root exudates are a complex material composed of polysaccharides, proteins, phospholipids, cells that detach from the external layers of the root and many other compounds. Respired CO_2 and organic acid anions that naturally decompose and can react with calcium-rich silicates within soils are of particular interest. David Manning (Newcastle University) and others [12] have documented that this process occurs naturally in soils at brownfields sites [13]. Cement and iron/steel production produces about 10 per cent of the world's carbon emissions. As proposed in figure 2, it may be possible to re-sequester a portion of this carbon into the subsurface through the admixing of recycled concrete and furnace slag into soils used for non-food crops and urban re-vegetation [14].

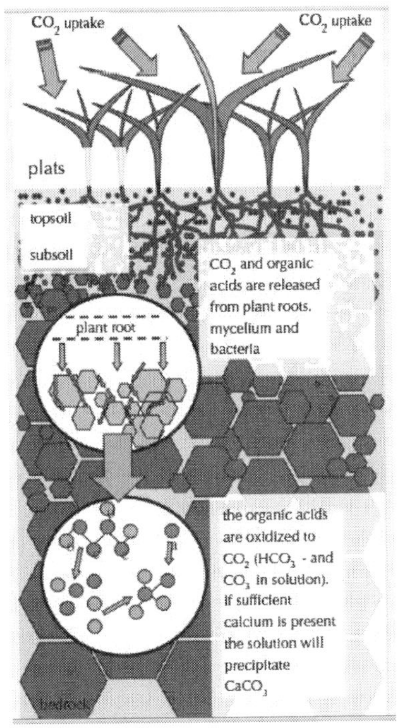

Figure 2: Overview of carbon sequestration through plant roots [14].

Stimulating and controlling naturally occurring biological processes provide an innovative solution for remediation of metal-contaminated ground water and soil *in situ* (e.g. [15,16]). Robert Smith (University of Idaho) and colleagues are attempting to co-precipitate radioactive heavy metals—such as strontium 90 (^{90}Sr), a uranium fission byproduct and carcinogen—with calcite, effectively immobilizing ^{90}Sr in solid form until it decays into innocuous isotopes (e.g. [17]). This requires immobilization on timescales long with respect to ^{90}Sr's half-life of 28.9 years. Laboratory (e.g. [18,19]) and modelling [20] studies indicate strong potential, and the technology has been considered for treatment at the US Department of Energy Hanford Site [21,22].

Successful water management and power generation around the world hinge largely on the retention and flow control of water in reservoirs and canals. In CA, USA alone more than 5000 km of levees and 1200 dams exist [23]. Construction quality ranges from fully to non-engineered, and their remediation is practically and fiscally prohibitive with conventional techniques. Yet, dam and levee breaches are often catastrophic to the society (e.g. Jones Tract, CA [24], Hurricane Katrina, New Orleans [25]). Researchers at Deltares, the Netherlands, recently completed field studies, wherein biofilms are used to reduce water seepage [26]. The procedure consists of injecting or stimulating biofilm-forming bacteria upstream of the seepage zone, the biofilm forces clogging of the soil matrix and the flow is reduced [27].

Other processes and applications with potential have been identified [28–30] and current research includes:

- biogeneration of N_2 gas *in situ* to desaturate soil for increased resistance to liquefaction [31];
- stabilization of surface soils against erosion [32–34];
- strengthening and repair of stone structures and monuments [35–38];
- rehabilitation of concrete [36,39–43];
- improved capacity of shallow foundations [44];
- dust suppression through agglomeration of fine particles [45];
- modification of ground water flow [46,47];
- biocementation of sands for liquefaction resistance [48];
- co-precipitation of heavy metals *in situ* [16,18].

These applications all focus on a single functionality or design objective, and in most cases research is in the early stages.

A multi-functional design that persists over generational timescales is the ultimate sustainable objective as the underlying biogeochemical processes are common among many applications. This represents a step change in how broad civil engineering problems are characterized and solutions are developed. A multi-functional application should seek to address multiple site-specific and regional environmental, infrastructure and agricultural demands and issues. A conceptual example for a motorway embankment is shown in figure 3. Using soil engineering *in vivo*, the soil embankment could become a multi-functional infrastructure component, simultaneously (i) reinforcing soils by biocementing using bacteria to generate mineral precipitates that support the roadway, railway and slope; (ii) stabilizing surface soils with plants to increase slope stability and minimize surface erosion; (iii) controlling surface and ground water flow by manipulating hydrogeological properties of soils (increase by biomineral dissolution and decrease by bioclogging); (iv) decontaminating surface water infiltrating the soil by bioremediation; and (v) capturing carbon dioxide by *in situ* inorganic precipitation of carbonate minerals.

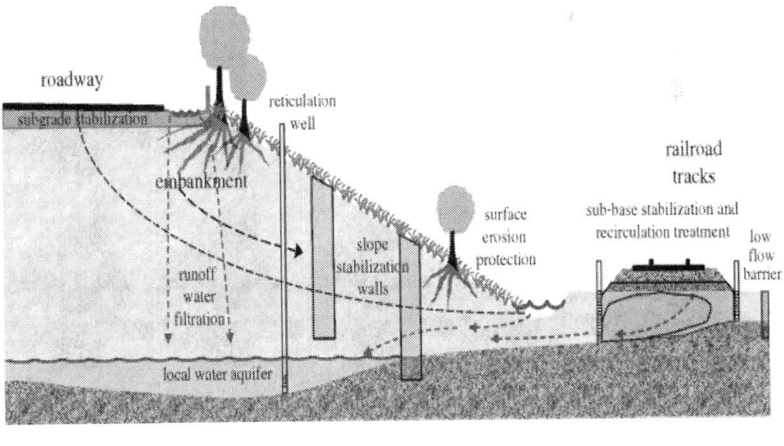

Figure 3: Schematic of multi-function solution using 'soil engineering *in vivo*'.

MULTI-FUNCTIONAL SYSTEMS: A FRAMEWORK FOR SOIL ENGINEERING IN VIVO

Soil engineering *in vivo* uses biogeochemistry (a combination of chemical reaction networks and biological processes) within the soil habitat. It also requires continuous process monitoring of this living system and upscaling to develop full-scale implementation procedures. The inter-relationship and interplay between these different components are presented separately for clarity in the schematic in figure 4 and subsequently, although they are highly interdependent.

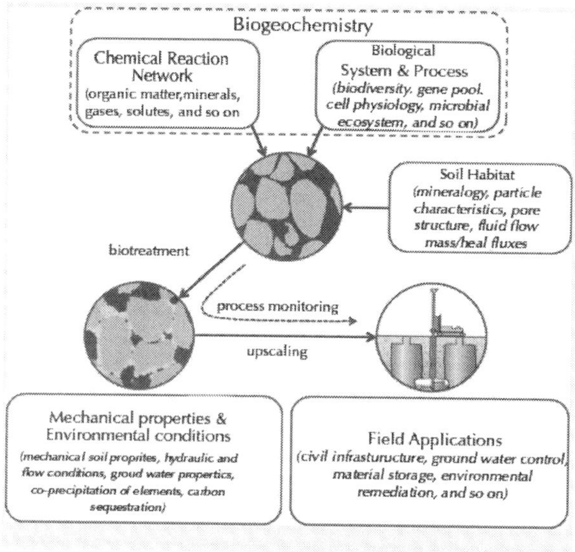

Figure 4: Framework for generalized approach to 'soil engineering *in vivo*'.

Biogeochemistry: Chemical Reaction Networks and Biological Systems and Processes

The biogeochemical framework for soil engineering in its broadest sense here, is directly taken from the current approaches used in bioremediation engineering; outlined in this section (e.g. [49]). The chemical reaction network and the biological systems and processes are inextricably linked in the near surface (<25 m), and herein are considered holistically as biogeochemistry [50]. Geochemical processes are driven by reactions between the elements that comprise soil organic matter, minerals, gases and solutes. One dominant condition for chemical reactions is the redox gradient that exists between the atmospheric $O_2(g)$ pool and its depletion in the subsurface owing to the influx and microbial decomposition of organic matter from the above-ground biomass [51]. When anoxia occurs below the depth of O_2 penetration, anaerobic processes degrade organic material, whereby respiration occurs using terminal electron acceptors other than O_2; notably, nitrate, higher valent manganese and iron minerals and sulphate [52]. Because hydrogen ion transfers accompany electron transfers to maintain charge balance in aqueous soil solutions, oxic environments (shallower) are relatively more acidic than where anaerobic processes dominate (deeper zones; [51]). Conversely, microbial respiration in these deeper strata—based on nitrate, sulphate and oxidized iron or manganese—produces net alkalinity [53]. A further condition driving chemical reactions is the inherent free energy of the parent material from rock mass, which formed under geological conditions of higher temperature and pressure than the ambient soil environment. Because most parent material is made up of complex silicate oxide minerals, their chemical decomposition to solutes and secondary minerals (e.g. clays) consumes hydrogen ions, thus contributing to the subsurface alkalinity.

The electron and hydrogen ion transfers between the various element pools control the master variables of pH and redox status. These dictate the chemical form of many metal compounds, whether dissolved, sorbed or precipitated, and thus determine their bioavailability, mechanisms of toxicity, and transport and mobility within the hydrologic cycle [54]. Furthermore, pH and redox status are major controls on rates of abiotic chemical reactions and set physiological limits for the soil as a habitat for microbial growth and activity. This is with respect to microbial ecosystem dynamics and associated microbial transformations of major and trace elements and organic contaminants [55,56].

Soil Habitat

The inherent complexity of soil deposits is because of the deposition and erosion of mineralogically complex parent material under natural transport processes, also the chemical complexity of soil organic matter, the largely unknown gene pool and physiology of the indigenous microbial populations, and their continuous interaction and dynamic evolution. Herein, the term 'soil habitat' is comprehensive, including the soil particle matrix, its mineralogy and associated microbial populations and the pore fluids and their composition. Critical attributes are wide-ranging for each component. For the soil matrix, important attributes include spatial variability, soil particle and aggregate size distribution, particle shape, mineral chemistry and surface charge, organic carbon characteristics and microscopic (immobile domain) porosity, as well as sessile microbial populations, and so on. For the pore fluid(s), important attributes are the physical properties governing flow such as density, viscosity, temperature, surface tension and pressure, as well as the attributes governing biogeochemical transformations (e.g. chemical composition, planktonic bacterial populations, suspended colloidal phases). Pore gas attributes include composition, pressure, temperature, and other physical and chemical characteristics. Pore gas may serve as a useful vehicle for controlling/removing volatile non-aqueous phases (via soil–vapour extraction schemes for instance), or in conveying water in vapour phase in desert environments. However, these complex and varied phenomena are often glossed over or ignored when considering saturated (ground water) zone manipulations and engineering, due in large part to the challenges in characterizing the subsurface variability, especially the microbial populations either naturally occurring (e.g. [57]) or under engineering control strategies (e.g. [58]).

Compatibility of Systems

Manipulation of a specific biogeochemical pathway (e.g. by amendment of soil with nutrients, other chemicals and biota such as plant crops and potentially microbes) for a desired outcome (e.g. calcite precipitation) can require/force a shift in the ambient environment. Physical and biogeochemical compatibility of the treatment with the ambient environment must be considered.

Physical compatibility is necessary for delivery of nutrients and/or biota uniformly throughout the subsurface. Identification of the limits of treatment effectiveness can be identified considering micro-organism size, pore throat size within the soil matrix, ambient stress conditions and treatment objective [30,48,59,60]. As shown in figure 5, micro-organisms may be motile and unaffected by ambient stress conditions in soils with particle sizes larger than about 2 μm. However, in smaller particle soils, the micro-organisms are probably trapped with diffusive nutrient transport becoming increasingly important [60].

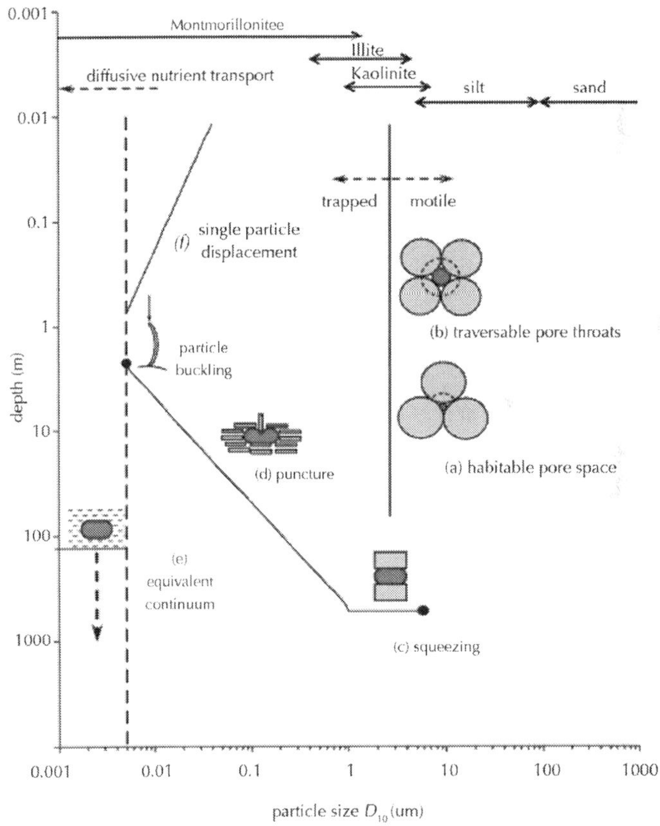

Figure 5: Overview of compatibility regimes considering particle size and ambient stress conditions (Rebata-Landa & Santamarina 2006. ©Copyright 2006 American Geophysical Union. Adapted from American Geophysical Union).

Competition among other natural biological species must be similarly examined when nutrients, other chemicals or new organisms are added to create a dominant biogeochemical pathway. In the natural environment, nutrients added for biostimulation are available to all systems, requiring the desired system to out-compete competitors and create an advantageous environment that must be sustained throughout a treatment process. Such competition among microbial guilds capable of mediating redox reactions with distinct (and respective per guild) electron-accepting chemical compounds is well established (e.g. [61]). Additionally, the role of protists as predators of subsurface microbial species (e.g. [62]) within the context of engineering stimulation of native species is not well studied or observed.

Process Monitoring

Integration of biogeochemical processes within soil requires continuous process monitoring of system indicators throughout the project service life. Markers to track the (in)equilibrium of the chemical reaction network and condition of the biological processes are well-documented in the bioremediation literature [52,63,64]. Shallow subsurface microbial communities capable of using distinct electron acceptors, often generate a 'redox-spectrum' along a flow path conveying nutrients sufficient for activity (e.g. along a contaminant plume path). The redox-spectrum appears as a continuous decrease in measured or theoretical electrochemical potential (see [52]) along the flow path, associated with sequential exhaustion of electron-accepting chemical compounds in the order of oxygen, nitrate, iron and manganese minerals, and sulphate. For the particular case of calcite precipitation, urea and its degradation product ammoni(um/a) can serve as indicators of reaction extent for this kinetically controlled reaction. Finally, the precipitation/dissolution potential for calcite is classically indicated by the saturation index. This parameter is zero when calcite is in equilibrium with the ground water; however, disequilibrium is often observed and can arise from and indicate incomplete mixing, lack of nucleation sites, pore-scale variation in pH, among other factors. Critical markers for soil processes depend on the application objective. Monitoring of chemical concentration in ground water provides direct assessment of water quality while nucleic acid-based analyses determine genetic and functional microbial diversity and measures of soil strength provide direct measures of improvement to mechanical soil properties.

In general, monitoring processes are either direct or indirect. Direct measurements typically are established, accurate and accepted standards that require discrete physical samples extracted from the subsurface, resulting in time-delayed measurements. Indirect measurements (e.g. electrical, magnetic and geophysical measurements) that provide real-time volume-averaged secondary indicator measurements are increasingly common [65,66]. The two methods are often used in tandem, with indirect measurements providing real-time monitoring and direct measurement providing occasional reference/calibration/verification data. It seems likely that with the development of bioengineered solutions, new approaches to monitoring will be developed, particularly expanding the use of biological activity beyond current bioremediation applications as a broader marker for soil engineering functions.

Upscaling for Field Application

Communal knowledge drops precipitously as processes are upscaled, and as a result upscaling is the most challenging issue for many soil engineering systems being researched. Upscaling approaches are multi-faceted, but often lead with experimental approaches enhanced by analytical and numerical work. Convention builds on examination of a biogeochemical process under stringent controlled conditions (i.e. sterile environment with select bacteria species and chemical/nutrients), and then relaxation of the conditions as it is applied to zero-dimensional batch reactors, one-dimensional soil columns, two-dimensional bench models, and then three-dimensional prototypes and full-scale pilot application. Monitoring complexity increases factorially with the increase in experimental variables. Modelling also parallels this increase in complexity, beginning with analytical modelling of specific biogeochemical variables in controlled batch or column systems, and then using biogeochemical codes linked with higher dimension flow and transport-simulation capabilities (e.g. TOUGHREACT, STOMP; see [56]) to simulate *in situ* conditions.

Upscaling towards *in situ* conditions with less control and observability involves a loss of information that changes the nature of the problem [67]. This is acutely demonstrated with heterogeneities in the velocity and concentration fields (due to non-uniform boundary conditions, stratigraphic spatial variability, subsurface reaction properties), which can produce complex solute distributions with sharp gradients and preferential flow and transport pathways [68]. These gradients are difficult to detect and yet can be a primary control on biogeochemical transformations. This has led to upscaling techniques that focus on Lagrangean measures (e.g. travel-time, cumulative reactivity) instead of conventional (Eulerian) plume delineation (e.g. [69]).

EMERGING TECHNOLOGY EXAMPLE: MICROBIALLY INDUCED CALCITE PRECIPITATION

One emerging process that holds promise for multi-functional sustainable applications is the biogeochemical process of carbonate precipitation. Microbially induced calcite precipitation (MICP) has many potential applications, including improvements to mechanical (e.g. strength, liquefaction resistance, stiffness) and hydraulic (e.g. reduction in permeability) soil properties, immobilization of heavy metals (e.g. ^{90}Sr and others) in the subsurface, and sequestration of atmospheric carbon into the subsurface [28,59]. MICP has been previously explored for microbially enhanced oil recovery [70,71] and builds on prior research on chemically induced calcite precipitation [72,73]. The details of this process are relatively well understood and although many scientific issues remain challenging to quantify, upscaling to field deployment has nearly arrived. This technology is presented herein to exemplify the framework and approach proposed above (figure 4).

Enabling Biogeochemical Processes

MICP occurs following [41]:

$$Ca^{2} + +HCO_3^- + OH^- \rightarrow CaCO_3 \downarrow + H_2O \qquad (4.1)$$

$$Ca^2 + +2HCO_3^- \leftrightarrow CaCO_3 \downarrow +CO_2 + H_2O \qquad (4.2)$$

The creation of calcium carbonate (calcite) occurs as a consequence of bacterial metabolic activity. When the rates of urea hydrolysis and calcite precipitation are equal, such that NH_4^+ cations effectively replace precipitating Ca_2^+ cations then the pH remains neutral [74]. When urea hydrolysis outpaces calcite precipitation, the pH of the proximal environment increases (e.g. [75]). Calcite precipitation may be achieved by many different processes [59], including urea hydrolysis, denitrification [76–78], sulphate reduction ([79], inducing dolomite precipitation) and iron reduction ([80], inducing ankerite and other mixed mineral precipitation). Enzymatic hydrolysis of urea by microbes is the most energy-efficient [59], and urease activity is found in a wide range of micro-organisms and plants [81]. Selection of the process ultimately implemented in the field should consider the differences in efficiency rates, contaminant byproducts and energy consumption between these processes. *Bacillus pasteurii* (American Type Culture Collection (ATCC) 6453) and a more recent strain, *Sporosarcina pasteurii* (ATCC 11859), an alkalophilic bacterium with a highly active urease enzyme [46], have been used in laboratory studies.

Systems Compatibility

Effective treatment with MICP depends on the compatibility between groups of soil components, examples of which are highlighted herein. Geometric compatibility between the microbe size (*S. pasteurii* are rod-shaped and approx. 1 μm in diameter and approx. 3–5 μm in length) and the soil structure pore size is necessary for delivery of micro-organisms for uniform treatment. Data have shown MICP treatment to be effective for soils with particle sizes of about 25 μm diameter (silty soils) and larger [31,82]. Biochemical compatibility is critical for controlling the rate and timing at which MICP occurs, and porewater compositions affect this. MICP increases in efficiency (assessed in terms of the rate at which soil stiffness increases during treatment) as the ground water salinity increases towards an oceanwater composition [82]. The initiation of precipitation is determined by treatment solution buffering, and appropriate proportions of ammonium chloride and sodium bicarbonate can control the onset of precipitation [83]. Soil particle mineralogy and surface characteristics also affect MICP. Calcium, silica and iron base mineral sands can all be effectively treated [82]. All natural sands (and silts) tested that are geometrically (size) compatible show improvements in soil stiffness with sufficient time.

Process Monitoring of Experiments

MICP process monitoring is exemplified through results from one-dimensional column experiments (length = 45 cm, diameter = 5 cm) performed on sand (D_{50} = 0.21 mm) to improve spatial uniformity (figure 6*a*) [84]. Similar approaches have been reported by Whiffin *et al*. [85]. The data obtained were indirect real-time measurements (e.g. pH, versus, optical density) and more accurate, time-delayed direct measurements (chemical concentrations, bacterial plate counts) from four column elevations throughout treatment. Following injection and throughout treatment, optical density (indirect) and plate counts (direct for viable cells) were performed on biotic samples to assess bacterial distribution (figure 6*b*). Calcite precipitation at particle–particle contacts was monitored in real time as changes in soil stiffness with (indirect) geophysical shear wave velocity measurements [86,87] (figure 6*c*). The shear wave velocity (normalized by the initial shear wave velocity at each monitoring point) increased at all measurement locations throughout treatment (figure 6*c*). Post-test measurements of precipitated calcite (direct) validate geophysical measurements of changes in soil stiffness (figure 6*d*).

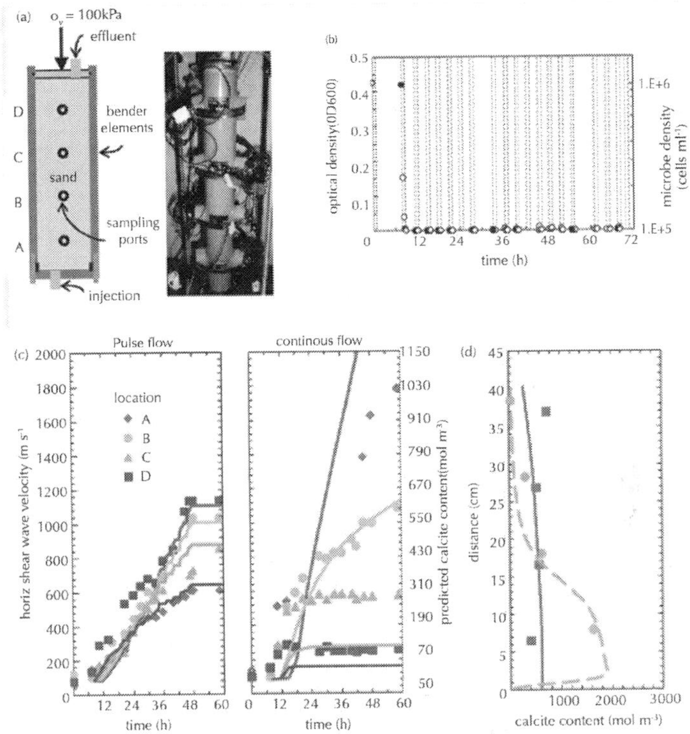

Figure 6: (*a*) One-dimensional column configuration, (*b*) direct (plate count, filled circles) and indirect (optical density, OD600, open circles) measurements throughout treatment, (*c*) evolution of experimental shear wave velocity and model-predicted calcite concentration with treatment time and, (*d*) agreement between experimental measures and numerical predictions of calcite distribution through column after treatment. Filled squares, pulse flow measured; filled circles, continuous flow measured; blue line, pulse flow predicted; green dotted line, continuous flow predicted.

Parallel Modelling of Experiments

Biogeochemical modelling of experimental MICP work at the batch-scale has confirmed the conceptual model underlying the general reaction network and its behaviour, and is enabling scoping assessment of more complex systems. Building on previous MICP analysis (e.g. [17,88]), the MICP reaction network was incorporated into the modelling framework TOUGHREACT [89–91], a multiphase, multi-component, reactive transport model [84]. The *minteq.v4.dat* thermodynamic database [92] and the kinetic expression for jack bean (*Canavalia ensoformis*)-derived enzymatic ureolysis [93] were adopted. Urease was simulated as an immobile species, and the spatial urease distribution was modelled as a log-linear relationship for biocolloid advection and filtration through an inverse solution using UCODE [94]. Even with these limitations, insights can be gained. Very good agreement in the spatial distribution of calcite is obtained by the model. With precipitated calcite concentration serving as a proxy for soil stiffness (as the relationship between the two is proportional; [95]), good agreement is also obtained in trends of shear wave velocity (soil stiffness) throughout the treatment process at all sampling locations along the test column (figure 6*c*,*d*).

EXPERIMENTS AND MODELLING AT SEPARATE SCALES

To date, MICP research has focused on centimetre and metre length scales under zero-delayed (batch) and one-dimensional (column) experiments. Insights gained at these laboratory scales are critical for full-scale implementation.

μm: Calcite Crystal Structure, Microbial Embedment

Scanning electron microscopy (SEM), laser-scanning confocal microscopy and energy-dispersive X-ray spectroscopy have revealed characteristics of the calcite crystal structure and the role of micro-organisms. SEM images have captured microbial 'beds' within the calcite structure, verifying that micro-organisms create local environmental changes so that cells serve as preferential calcite nucleation sites (figure 7*a*). Variations of the soil type and treatment formulation also alter the calcite structure; the conventional treatment formula ([15,48,75]) produces a random amorphous calcite structure.

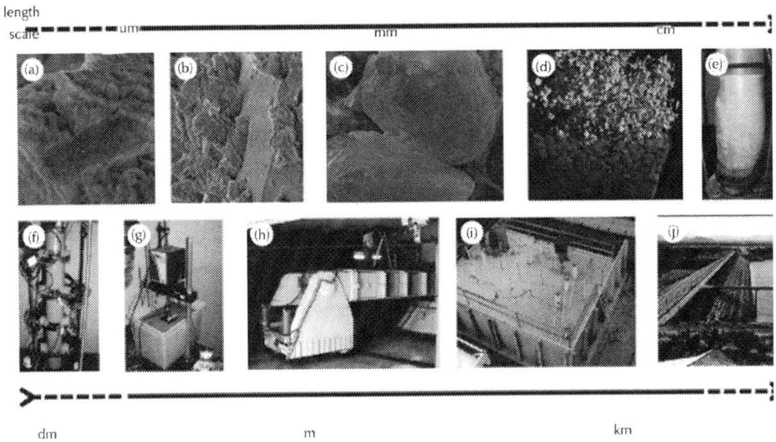

Figure 7: Overview of experimental upscaling of MICP (*a*) microbe impression in precipitated calcite, (*b*) amorphous calcite on silica sand particle surface, (*c*) MICP cementation of two sand particles, (*d*) X-ray CT scan of MICP-treated sand, (*e*) triaxial element laboratory testing, (*f*) one-dimensional column experiments, (*g*) 1-G model testing of shallow foundation, (*h*) centrifuge testing of systems supported on MICP, (*i*) MICP treatment of 100 m^3 sand [78] and (*j*) possible structure for field implementation.

mm: Particle–particle Cementation and Degradation, Spatial Calcite Distribution

SEM and X-ray computed tomography (CT) at the particle level capture the spatial distribution of calcite. MICP has apparent preference towards particle–particle contacts owing to bacterial filtration and greater access to nutrients available in the flowing pore fluid (figure 7*b*,*c*). Calcite crystal thicknesses on exposed particle surfaces are less than approximately 10 μm, while structures at particle–particle contacts is much higher for saturated sands. Unsaturated sands treated by percolation results in calcite precipitation almost exclusively at particle contacts. Degradation of particle–particle contacts typically occurs through the calcite structure and not at the calcite–silica interface. Three-dimensional views from CT scanning confirm SEM observations and reveal that the pore structure remains connected after treatment (figure 7*d*) [59].

cm: Improvements to Engineering Properties, Monitoring Technique Developments

Element test results on MICP-treated sand have revealed the potential improvement to engineering, mechanical and hydraulic properties and enabled the development of real-time monitoring techniques (figure 7e). MICP treatment can reduce hydraulic conductivity [46,85], increase stiffness such that loose sand becomes soft rock [59,77,78] and alter volumetric behaviour [59]. The universality of MICP has been examined using different soil gradations, mineralogies and ground water conditions. These tests have also enabled improvement of geophysical measurements [95–97] and discrete sampling protocols necessary at larger length scales.

dm: One-Dimensional Columns for Treatment Uniformity, Bench-Scale Models Assessing Improved Geosystem Performance

Extension of element specimens in one-dimensional (figure 7*f*) to multi-dimensional models (figure 7*g*) enables treatment optimization and insight at the 1-G model scale of how geosystem performance (e.g. foundations) may be improved with MICP. MICP has been successfully implemented in one-dimensional columns to a distance of 5 m [85], and shorter 0.5 m one-dimensional columns have been used to improve the timing, rate and uniformity of MICP treatment (per above data [84]). Martinez & DeJong [44] demonstrated using 1-G model tests that the capacity of shallow foundations could be increased fivefold or more through cementation of near-surface sands.

m and km: Three-dimensional Test Chambers, Centrifuge Modelling and Field Scale Treatment

Initial results are now being generated for three-dimensional test chambers and field conditions. Centrifuge modelling is being implemented to assess the performance of MICP-treated soil and supported structures when subjected to dynamic earthquake loading. Following successful treatment of 1 m^3 soil, treatment of a 100 m^3 soil specimen has been performed (figure 7*i*) [77,78].

Limited progress has been made in upscaling MICP process models for general use, in part, as a result of the strong and complex coupling among flow, mixing and reaction (e.g. [98]). However, for particular engineering applications at field-scale, research at all length scales continues as improved scientific understanding is continually sought. Specific challenges and issues that remain include precise understanding and modelling of bacterial distributions in the natural subsurface and associated ureolysis rates, exploration of alternative processes such as denitrification, mapping subsurface stratigraphy, flow and transport through heterogeneous media, durability and permanence of treatment, and biodiversity and behaviour in nutrient-limited environments.

The primary research mindset and focus has been on single function applications (with different research groups around the world examining MICP for different applications). The multi-function application proposed herein requires additional scientific and engineering research before the performance of MICP can be predicted and executed in the complex natural environment.

MULTI-FUNCTIONAL SOLUTIONS FROM FIRST PRINCIPLES: CHALLENGES AHEAD

Multi-functional solutions (e.g. figure 3) using MICP in combination with other processes cannot be realized using empirically based and conservative design methods; a step change in our design approach, wherein analysis of these complex systems based on first principles is needed. In the nineteenth century, structural engineers began designing based on the first principles of solid mechanics. In the twentieth century, water treatment engineering incorporated aqueous chemistry in an analogous manner [99]. The interface of the twenty-first century is between biology and civil and environmental engineering, and requires a similar step change in theoretical concepts for application, engineering analysis and design principles.

It is daunting to consider the application of biogeochemical engineering processes in natural environments. Some of the key biological challenges, for example, are the predominance of unknown genotype and phenotype in the subsurface, and the complexity of indigenous microbial consortia and their ecological function. Fluid movement and mass transport pathways depend on spatially variable physical properties, such as permeability that are not easily mapped from the surface. Lithological structures yield spatially variable mineralogy with attendant differences in mineral surface hydrophobicity, charge, chemical bonding environment and trace element provenance for organisms.

Spatial variability is critical for biological processes. Biota respond to the physical and chemical environment, and ecosystem dynamics are influenced by the temporal variability of flow and transport within a spatially variable environment. Microbial function is dominated by attached growth communities in the subsurface, and the spatially variable mineralogy is likely to exert strong selective pressure and impact community structure and function. A key task will be to understand how different biological processes interact. For example, soil stabilized with MICP is likely to be covered with vegetation. It will be important to establish that activity at the root scale has a synergistic (and not antagonistic) effect on MICP. The ability of mycorrhizae associated with plant roots to release organic acids that result in mineral dissolution underscore these concerns. Such considerations will be novel challenges to engineers who may presently be comfortable with designing structures of concrete and steel, which are assumed to be spatially and temporally invariant with respect to material parameters.

These challenges are evident in the current engineering practice for harnessing subsurface biological processes (e.g. land and ground water remediation). Extensive site-specific data are required to infer, for example, solute transport and biodegradation rates. Current advances in computational methods and site investigation allow numerical models to consider a great deal of this complexity; on a computational scale which approaches that of, for example, climate simulations [63]. However, predictions obtained are largely based on simple extrapolation of prior results from laboratory/bench scales to larger length and timescales. This extrapolation can be fraught with problems as, for example, highly localized mixing-limiting processes at the micro-scale are captured in equivocal volume-averaged models that mask critical process-limiting factors (e.g. [100]). Further, little is known about how the system will evolve spatially and temporally, and there is no fundamental approach that provides parameter values for desktop analysis prior to expensive, site-specific engineering investigation.

To address these challenges, it is necessary to establish the theoretical framework and tools to assess the subsurface biogeochemical-soil habitat and the subsurface spatial variability of the respective components. This framework must then be applied to research experimental design, to target the hypotheses drawn from the above challenges, target data acquisition to address these specific challenges and develop improved methods for *in situ*observation. This begins to point to a research agenda to develop analysis and design methods that are based on first principles: (i) sound biological theory; (ii) hypothesis-driven field experimentation; and (iii) new sensing methods that are preferably non-invasive and non-destructive.

SUSTAINABILITY: COMPREHENSIVE DESIGN CRITERIA

Multi-functional solutions based on first principles will be evaluated against the broader social criterion that is emerging. In the twentieth century, energy-intensive physical practices (e.g. cement manufacturing, *ex situ* soil remediation, strip mining, and so on) continued for decades owing to perceived abundance of natural resources and unawareness of the consequences. Today, it is clear that the resources are finite and the impact and costs of past consumption are compounding (e.g. [101,102]).

The twenty-first century will be characterized as the era of sustainability [103]. The historical project assessment based solely on financial investment is incomplete. Projects must now be evaluated against the social obligation we have to limit environmental impact, an holistic assessment of energy embodied in the project, the project's carbon and water footprint, the probabilities of failure and rebuilding throughout the project service life, and the ability for material and site reuse for future generations [104,105].

Many twentieth century technologies are inefficient within this sustainability criterion. To an extent possible, they must be replaced by new technologies, and 'soil engineering *in vivo*' for multi-functional solutions is proposed to be one of them. Conceptually, the harnessing of natural biogeochemical processes for multi-functional solutions embodies the sustainability criteria; ideally, it would consume less energy, perform treatment *in situ*, generate minimal waste, minimize hazards to humans and the environment, leave the soil available for a symbiotic use (e.g. available for filtration of ground water while supporting a building) or a subsequent reuse (e.g. available for use as structural foundation following site remediation), alter/manipulate environment to the minimum extent needed, and so on.

Evaluation of the different biogeochemical solutions currently being researched must be evaluated against this broader criterion. Improvement relative to conventional twentieth century practice for a specific application is desired (and would be certainly welcome), but even further societal benefits gained may be realized through the multi-functional approach proposed herein.

INSTITUTIONAL CHALLENGES: REFORMATION IS NECESSARY

Realization of these multi-functional solutions in society necessitates a step change in the culture and ethos of research, education and policy.

Research Reform

This research is inherently interdisciplinary; scientists and engineers must become intertwined. Change is required at the individual and institutional levels. Scientists must move towards more proactive engagement, control and manipulation of biological process so that biological processes can be reliably harnessed in engineering applications. Simultaneously, engineers must consider soil (and the systems built upon it) as living ecosystems. The mindset of once-and-done design and construction must be transformed into continuous performance monitoring of infrastructure. Fortuitously, health monitoring of infrastructure (e.g. [106–108]), which is becoming more widely accepted, is the first step. At the institutional level, this new research may not be amenable to the conventional organization and culture in which specialties such as civil engineering remain compartmentalized. Interdisciplinary entities/groups within which ideas can be generated, vetted, proofed, explored and developed freely are necessary. The culture of these groups and the broader institution must value these activities.

Education Reform

The growth and maturation of this new field over the coming decades require development of new engineers and scientists. Education—and (post-) graduate education in particular—in the latter part of the twentieth century that focused on individuals developing expertise in a single niche area is insufficient for this interdisciplinary field. Individuals must have expertise (core competency) in a certain area(s), but must be knowledgeable and able to communicate across several different disciplines [59]. In essence, a seventeenth century type renaissance education model is necessary. Creativity, innovation, mind experiments and critical thinking at the systems level must be emphasized. Fortunately, new education modules such as interdisciplinary seminar series, interdisciplinary (co-taught) courses, interdisciplinary graduate programmes and graduate programmes requiring a minor in another discipline, science/engineering/business combined with Master's degrees, and doctoral training degrees (interdisciplinary coursework only degree) are emerging [28].

Industry and Government Restructuring

The movement towards sustainable performance-based design and the associated changes beginning in construction policies and practice must continue. Cost assessment must be expanded beyond the physical footprint and construction cost to include life-cycle analysis of all components, ranging from the manufacturing of raw materials through to site re-use feasibility after the service life expires. This may also require government subsidies/credits (similar to that currently being given for renewable energy sources) as a comprehensive sustainable assessment (cradle to grave) of materials and construction may (or will likely) lead to a final design that is more expensive when analysed under the current costing/bidding framework. This may lead to government subsidies for private projects and the restructuring of public work projects, where the design selection criteria could be changed/modified to require/endorse consideration of a broader set of factors in a sustainable framework.

CONCLUSIONS

Society's major civil and environmental challenges cannot be solved using the intensive energy twentieth century treatment solutions. New sustainable solutions are needed. It is proposed herein that 'soil engineering *in vivo*', an emerging interdisciplinary approach which recognizes soil as a living, active system that has the capacity to provide sustainable solutions, may be one of them.

The envisioned applications are many and wide-ranging, from anti-desertification measures for northern Africa to reducing the liquefaction settlement for a foundation building. To date, applications investigated are largely single function solutions, but herein it is proposed that select biogeochemical systems can simultaneously provide multi-functional solutions.

Multi-functional solutions must consider the processes occurring within soil systems both independently and in a coupled context. Techniques for real-time monitoring of these integrative solutions are critical. Upscaling requires extensive experimental work that is hypothesis-driven owing to the inherent complexity of natural systems. The experimental work must be complemented by rigorous analytical and numerical modelling to identify the first-order principles controlling the system.

MICP has rapidly matured towards field implementation, with potential applications including improvements to mechanical and hydraulic soil properties, immobilization of heavy metals and sequestration of atmospheric carbon. Experimental and numerical modelling, including real-time process monitoring, have provided significant advances in understanding and control of MICP at all length scales.

Realization of MICP and similar processes as field-scale multi-functional solutions requires a step change in how science and engineering is performed, and must meet the emerging sustainability criterion. Significant additional research and development is needed in order to provide solutions within the natural environmental that are based on first principles. The primary challenges in this area include assessment of subsurface ecosystems and their interaction with the available fluids and minerals, and mapping and modelling of subsurface variability of pore structure, mineralogy, ground water flow and transport, and bacterial diversity.

Finally, realization of these multi-functional solutions in society necessitates a step change in the culture and ethos of research, education and policy. Interdisciplinary research requires greater interaction between scientists and engineers, formation of interdisciplinary work groups and departments, and institutional embrace of interdisciplinary work. Maturation of this field requires a new generation of scientist and engineers with a 'renaissance type' education equipped to use their own expertise and leverage the knowledge in interdisciplinary teams. Industry and government reform is needed to fund the development of new solutions, and incentives and evaluation criteria that encourage their deployment through industry should be developed.

ACKNOWLEDGEMENTS

Funding provided by the United States National Science Foundation (nos 0727463 and 0628782), Geosyntec Inc. and the UC Discovery Grant Programme in support of the research by J.T.D., D.C.N. and their group. Any opinions, findings and conclusions or recommendations expressed in this material are those of the writer(s) and do not necessarily reflect the views of the National Science Foundation. Mr Michael Fritzges, Mr Matthew Weil, Mr Jack Waller collaborated in their research. T.G. and T.B. were supported by the Office of Science (Biological and Environmental Research), US Department of Energy, grant no. DE-FG02-07ER64404. K.S., S.A.B. and W.R.W. were supported by the UK Engineering and Physical Science Research Council (EP/E031935/1). The contributions of all the participants of the 2007 International Workshop on BioSoils Interactions and Engineering for their ideas regarding possible applications and the future research and education needs are acknowledged and appreciated. Rothamsted Research is an institute of the UK Biotechnology and Biological Sciences Research Council.

REFERENCES

1. IPCC 2005. Carbon dioxide capture and storage. Spain: Intergovernmental Panel on Climate Change
2. USEPA 2008. Green remediation: incorporating sustainable environmental practices into remediation of contaminated sites. US Environmental Protection Agency, Office of Solid Waste and Emergency Response. http://www.clu-in.org/download/remed/Green-Remediation-Primer.pdf
3. Commission of the European Communities 2006. *Thematic strategy for soil protection*. Communication from the Commission to the Council, the European Parliament, The European Economic and Social Committee and the Committee of the Regions. COM(2006)241 final, 12 pp
4. Donnelley R. R. 2009. The Scottish soil framework, Edinburgh: The Scottish Government
5. CEEQUAL 2008. The civil engineering environmental quality assessment and awards scheme manual, v. 4, pp. 114 CEEQUAL Ltd
6. LEED 2009. New construction and major renovations rating system. Washington, USA: US Green Building Council
7. Larsson M. 2009. Turning dunes into architecture. Seehttp://www.ted.com/talks/magnus_larsson_turning_dunes_into_architecture.html(accessed 14 April 2010)
8. United Nations University 2007. Earth Summit + 5; UN Conventional to Combat Desertification, DPI/1870/SD. Seehttp://www.un.org/ecosocdev/geninfo/sustdev/desert.htm, DPI/1870/SD
9. KuzyakovY., Domanski G. 2000. Carbon input by plants into the soil: review. J. Plant Nutr. Soil Sci. 163, 421–431 (doi:10.1002/1522-2624(200008)163:4<421::AID-JPLN421>3.0.CO;2-R)10.1002/1522-2624(200008)163:4<421::AID-JPLN421>3.0.CO;2-R

10. Taylor L. L., Leake J. R., Quirk J., Hardy K., Banwart S. A., Beerling D. J. 2009.Biological weathering and the long-term carbon cycle: integrating mycorrhizal evolution and function into the current paradigm, page 185. Geobiology 7, 171–191 (doi:10.1111/j.1472-4669.2009.00194.x)10.1111/j.1472-4669.2009.00194.x

11. Jones D. L., Nguyen C., Finlay R. D. 2009. Carbon flow in the rhizosphere: carbon trading at the soil–root interface. Plant Soil 321, 5–33 (doi:10.1007/s11104-009-9925-0)10.1007/s11104-009-9925-0

12. Sanderson K. 2008. Waste concrete could help to lock up carbon. Nature(doi:10.1038/news.2008.732)10.1038/news.2008.732

13. Renforth P., Manning D. A. C., Lopez-Capel E. 2009. Carbonate precipitation in artificial soils as a sink for atmospheric carbon dioxide. Appl. Geochem. 24, 1757–1764 (doi:10.1016/j.apgeochem.2009.05.005)10.1016/j.apgeochem.2009.05.005

14. Manning D. A. C. 2008. Biological enhancement of soil carbonate precipitation:passiveremovalofatmosphericCO_2.Mineralog.Mag. 72, 639–649 (doi:10.1180/minmag.2008.072.2.639)10.1180/minmag.2008.072.2.639

15. Fujita Y., Ferris F. G., Lawson R. D., Colwell F. S., Smith R. W. 2000. Calcium carbonate precipitation by ureolytic subsurface bacteria. Geomicrobiol. J. 17, 305–318 (doi:10.1080/782198884)10.1080/782198884

16. Fujita Y., Taylor J. L., Gresham T. T., Delwiche M., Colwell F., McLing T. L., Petzke L. M., Smith R. W. 2008. Stimulation of microbial urea hydrolysis in groundwater to enhance calcite precipitation. Environ. Sci. Technol. 42, 3025–3032 (doi:10.1021/es702643g)10.1021/es702643g

17. Fujita Y., Redden G. D., Ingram J. S., Cortez M. M., Smith R. W. 2004. Strontium incorporation into calcite generated by bacterial ureolysis. Geochim. Cosmochim. Acta68, 3261–3270 (doi:10.1016/j.gca.2003.12.018)10.1016/j.gca.2003.12.018

18. Dupraz S., Parmentier M., Ménez B., Guyot F. 2009. Experimental and numerical modeling of bacterially induced pH increase and calcite precipitation in saline aquifers.Chem. Geol. 265, 44–53 (doi:10.1016/j.chemgeo.2009.05.003)10.1016/j.chemgeo.2009.05.003

19. Ferris F. G., Phoenix V., Fujita Y., Smith R. W. 2004. Kinetics of calcite precipitation induced by ureolytic bacteria at 10 to 20°C in artificial groundwater. Geochim. Cosmochim. Acta 68, 1701–1710 (doi:10.1016/S0016-7037(03)00503-9)10.1016/S0016-7037(03)00503-9

20. Weathers T. S., Ginn T. R., Spycher N., Barkouki T. H., Fujita Y., Smith R. W. 2009. Modeling of ureolytic calcite precipitation for the remediation of Sr-90 using a variable velocity streamtube ensemble. Proc. of the American Geophysical Union, Fall Meeting 2009, 14–18 December 2009, San Francisco, CA. Abstract no. H41D-0933. Washington, DC: American Geophysical Union

21. Fujita Y., Taylor J. L., Wendt L., Reed D., Smith R. W. 2009. Potential for ureolytically driven calcite precipitation to remediate strontium-90 at the Hanford 100-N area. Proc. of the American Geophysical Union, Fall Meeting 2009, 14–18 December 2009, San Francisco, CA. Abstract no. H21H-07. Washington, DC: American Geophysical Union

22. Peterson R. E., Poston T. M. 2000. Strontium-90 at the Hanford site and its ecological implications. PNNL-13127 technical report. Washington, DC: Pacific Northwest National Laboratory (doi:10.2172/965213)10.2172/965213

23. DSOD 2010. Division of Safety of Dams, California Department of Water Resources. See http://www.water.ca.gov/damsafety/damlisting/index.cfm. (accessed 15 April 2010)

24. Mierzwa M., Suits B. 2005. Chapter 3: Jones Tract 2004 Levee Break DSM2 Simulation. Methodology for flow and salinity estimates in the Sacramento–San Joaquin Delta and Suisun Marsh 26th Annual Progress Report. CA, USA: California Department of Water Resources

25. Burton M. L., Hicks M. J. 2005. Hurricane Katrina: preliminary estimates of commercial and public sector damages. WV, USA: Center for Business and Economic Research, Marshall University

26. Blauw M., Lamber J. W. M., Latil M. N. 2009. Biosealing: a method for *in situ*sealing of leakages. In Ground improvement technologies and case histories (eds Leung C. F., Chu J., Shen R. F., editors.), pp. 125–130 Singapore: Geotechnical Society of Singapore (GeoSS)

27. Thullner M., Baveye P. 2008. Computational pore network modeling of the influence of biofilm permeability on bioclogging in porous media. Biotechnol. Bioeng. 99, 1337–1351 (doi:10.1002/bit.21708)10.1002/bit.21708

28. DeJong J. T., Mortensen B., Martinez B. 2007. *Bio-soils interdisciplinary science and engineering initiative*. NSF Final report for grant no. CMS-0628782, 85 pp

29. Ivanov V., Chu J. 2008. Applications of microorganisms to geotechnical engineering for bioclogging and biocementation of soil in situ. Rev. Environ. Sci. Biotechnol. 7, 139–153

30. Mitchell J. K., Santamarina J. C. 2005. Biological considerations in geotechnical engineering. ASCE J. Geotech. Geoenviron. Eng. 131, 1222–1233 (doi:10.1061/(ASCE)1090-0241(2005)131:10(1222))10.1061/(ASCE)1090-0241(2005)131:10(1222)

31. Rebata-Landa V. 2007. Microbial activity in sediments: effects on soil behavior. PhD thesis, Civil and Environmental Engineering, Georgia Institute of Technology, GA, USA

32. Fatahi B., Khabbaz H., Indraratna B. 2010. Bioengineering ground improvement considering root water uptake model. Ecol. Eng. 36, 222–229 (doi:10.1016/j.ecoleng.2008.12.027)10.1016/j.ecoleng.2008.12.027

33. Gray D. H., Sotir R. B. 1996. Biotechnical and soil bioengineering slope stabilization: a practical guide for erosion control. New York, NY: Wiley-Interscience

34. Loades K. W., Bengough A. G., Bransby M. F., Hallett P. D. 2010. Planting density influence on fibrous root reinforcement of soils. Ecol. Eng. 36, 276–284 (doi:10.1016/j.ecoleng.2009.02.005)10.1016/j.ecoleng.2009.02.005

35. Adolphe J. P., Loubière J. F., Paradas J., Soleilhavoup F. 1990. Procédé de traitement biologique d'une surface artificielle. European patent 90400G97.0. (After French patent 8903517, 1989)

36. De Muynck W., Verbeken K., De Belie N., Verstraete W. 2010. Influence of urea and calcium dosage on the effectiveness of bacterially induced carbonate precipitation on limestone. Ecol.

Eng. 36, 99–111 (doi:10.1016/j.ecoleng.2009.03.025)10.1016/j.ecoleng.2009.03.025

37. Le Metayer-Levrel G., Castanier S., Orial G., Loubiere J. F., Perthuisot J. P. 1999.Applications of bacterial carbonatogenesis to the protection and regeneration of limestones in buildings and historic patrimony. Sediment. Geol. 126, 25–34 (doi:10.1016/S0037-0738(99)00029-9)10.1016/S0037-0738(99)00029-9
38. Rodriguez-Navarro C., Rodriguez-Gallego M., Ben Chekroun K., González-Munoz M. T. 2003. Conservation of ornamental stone by *Myxococcus anthus*-induced carbonate biomineralization. Appl. Environ. Microbiol. 69, 2182–2193 (doi:10.1128/AEM.69.4.2182-2193.2003)10.1128/AEM.69.4.2182-2193.2003
39. De Muynck W., Debrouwer D., De Belie N., Verstraete W. 2008. Bacterial carbonate precipitation improves the durability of cementitious materials. Cement Concrete Res. 38, 1005–1014
40. Ramachandran S. K., Ramakrishnan V., Bang S. S. 2001. Remediation of concrete using micro-organisms. ACI Mater. J. 98, 3–9
41. Ramakrishnan V., Bang S. S., Deo K. S. 1998. A novel technique for repairing cracks in high performance concrete using bacteria. Proc. Int. Conf. on HPHSC, 10–12 August 1998, Perth, Australia, pp. 597–618
42. Ramakrishnan V., Ramesh K. P., Bang S. S. 2001. Bacterial concrete. Proc. SPIE4234, 168–176 (doi:10.1117/12.424404)10.1117/12.424404
43. De Muynck W., De Belie N., Verstraete W. 2010. Microbial carbonate precipitation in construction materials: a review. Ecol. Eng.36,118–136(doi:10.1016/j.ecoleng.2009.02.006)10.1016/j.ecoleng.2009.02.006
44. Martinez B. C., DeJong J. T. 2009. Bio-mediated soil improvement: load transfer mechanisms at the micro- and macro-scales. US–China Workshop on Ground Improvement Technologies. ASCE GSP 188, 242–251
45. Bang S. S., Bang S., Frutiger S., Nehl L. M., Comes B. L. 2009. Application of novel biological technique in dust suppression.

Transportation Research Board 88th Annual Meeting, *11–15 January 2009*. Washington, DC: Transportation Research Board

46. Ferris F. G., Stehmeier L. G., Kantzas A., Mourits F. M. 1996. Bacteriogenic mineral plugging. J. Can. Petrol. Technol. 35, 56–61
47. Nemati M., Voordouw G. 2003. Modification of porous media permeability, using calcium carbonate produced enzymatically *in situ*. Enzyme Microbial Technol. 33, 635–642 (doi:10.1016/S0141-0229(03)00191-1)10.1016/S0141-0229(03)00191-1
48. DeJong J. T., Fritzges M. B., Nüsslein K. 2006. Microbial induced cementation to control sand response to undrained shear. ASCE J. Geotech. Geoenviron. Eng. 132, 1381–1392 (doi:10.1061/(ASCE)1090-0241(2006)132:11(1381))10.1061/(ASCE)1090-0241(2006)132:11(1381)
49. Alvarez P. J. J., Illman W. A. 2006. Bioremediation and natural attenuation: process fundamentals and mathematical models. Chichester, UK: Wiley-Interscience
50. Banwart S. A., Thornton S. F. 2003. The geochemistry and hydrology of groundwater bioremediation by natural attenuation. In Bioremediation: a critical review(eds Head I., Singleton I., editors.), pp. 93–138 Horizon Scientific Press
51. Banwart S. A. 1996. Groundwater geochemistry in the burial environment. InPreserving archaeological remains in situ (eds Corfield M., Hinton P., Nixon T., Pollard M., editors.). Bradford, UK: Department of Archaeological Sciences, University of Bradford
52. Christensen T. H., Bjerg P., Banwart S. A., Jakobsen R., Heron G., Albrechtesen H.-J. 2000. Characterization of redox conditions in groundwater contaminant plumes. J. Contam. Hydrol. 45, 165–241 (doi:10.1016/S0169-7722(00)00109-1)10.1016/S0169-7722(00)00109-1
53. Jørgensen B. B. 2006. Bacteria and marine biogeochemistry. In Marine geochemistry(eds Schulz H. D., Zabel M., editors.), pp. 169–206 2nd edn Berlin, Germany: Springer
54. Younger P. L., Banwart S. A., Hedin R. 2002. Mine water. hydrology, pollution, remediation In Environmental pollution

series (eds Alloway B. J., Trevors J. T., editors.). Amsterdam, The Netherlands: Kluwer

55. Watson I., Oswald S. E., Mayer K. U., Wu Y., Banwart S. A. 2003. Modelling kinetic processes controlling hydrogen and acetate concentrations in an aquifer-derived microcosm. Environ. Sci. Technol. 37, 3910–3919 (doi:10.1021/es020242u)10.1021/es020242u
56. Watson I. A., Oswald S. E., Banwart S. A., Crouch R. S., Thornton S. F. 2005.Modelling the dynamics of fermentation and respiratory processes in a groundwater plume of phenolic contaminants interpreted from laboratory- to field-scale. Environ. Sci. Technol. 39, 8829–8839 (doi:10.1021/es0507970)10.1021/es0507970
57. Goldscheider N., Hunkeler D., Rossi P. 2006. Review: microbial biocenoses in pristine aquifers and an assessment of investigative methods. Hydrogeol. J. 14, 926–941 (doi:10.1007/s10040-005-0009-9)10.1007/s10040-005-0009-9
58. Balkwill D., et al. 2001. Breakthroughs in field-scale bacterial transport. EOS Trans. AGU 82, 417,423–425
59. DeJong J. T., Mortensen B. M., Martinez B. C., Nelson D. C. 2010. Bio-mediated soil improvement. Ecol. Eng. 36, 197–210 (doi:10.1016/j.ecoleng.2008.12.029)10.1016/j.ecoleng.2008.12.029
60. Rebata-Landa V., Santamarina J. C. 2006. Mechanical limits to microbial activity in deep sediments. Geochem. Geophys. Geosyst. 7, 12
61. Thullner M., Regnier P., Van Cappellen P. 2007. Modeling microbially induced carbon degradation in redox-stratified subsurface environments: concepts and open questions. Geomicrobiol. J. 24, 139–155 (doi:10.1080/01490450701459275)10.1080/01490450701459275
62. Novarino G., et al. 2006. Protistan communities in aquifers: a review. FEMS Microbiol. Rev. 20, 261–275 (doi:10.1111/j.1574-6976.1997.tb00313.x)10.1111/j.1574-6976.1997.tb00313.x
63. Banwart S. A., Thornton S. F. 2010. Natural attenuation of hydrocarbon compounds in groundwater. In Handbook of hydrocarbon and lipid microbiology (ed. Timmis K. N., editor.),

Berlin, Germany: Springer

64. Thornton S., Quigley S., Spence M., Banwart S., Bottrell S., Lerner D. 2001.Processes controlling the distribution and natural attenuation of phenolic compounds in a deep sandstone aquifer. J. Cont. Hydrol. 53, 233–267 (doi:10.1016/S0169-7722(01)00168-1)10.1016/S0169-7722(01)00168-1

65. Santamarina J. C., Klein K. A., Fam M. A. 2001. Soils and waves. Chichester, UK: John Wiley & Sons, ltd

66. Snieder R., et al. 2005. Advanced noninvasive geophysical monitoring techniques.Ann. Rev. Earth Planet. Sci. 35, 653–683 (doi:10.1146/annurev.earth.35.092006.145050)10.1146/annurev.earth.35.092006.145050

67. Cushman J. H., Benneethum L. S., Hu B. X. 2002. A primer on upscaling tools for porous media. Adv. Water Resour. 25, 8–12

68. MacQuarrie K. T. B., Sudicky E. 1990. Simulation of biodegradable organic contaminants in groundwater: 2. Plume behaviour in uniform and random flow fields.Water Resour. Res. 26, 223–239

69. Seeboonruang U., Ginn T. R. 2006. Upscaling aquifer reactivity via exposure-time concept: forward model. J. Contam. Hydrol. 84, 127–154 (doi:10.1016/j.jconhyd.2005.12.011)10.1016/j.jconhyd.2005.12.011

70. Clark J. B., Munnecke D. M., Jenneman G. E. 1981. *In situ* microbial enhancement of oil production. Dev. Industr. Microbiol. 22, 695–701

71. McInerney M. J., Jenneman G. E., Revus D. E., Knapp R. M., Menzie D. E. 1982.Selectivity of microbial plugging processes for enhanced oil-recovery. Abstracts of Papers of the American Chemical Society, 184, 32-INDE, Washington, DC: American Chemical Society

72. Ismail M. A., Joer H. A., Randolph M. F., Meritt A. 2002. Cementation of porous materials using calcite. Géotechnique 52, 313–324

73. Kucharski E. S., Price G. P., Li H., Joer H. A. 1996. Laboratory evaluation of CIPS cemented calcareous and silica sands. Proc. of 7th Australia New Zealand Conf. on Geomechanics, Adelaide, Australia. pp. 102–107 Barton, Australia: Institution of Engineers

74. van Paassen L. A. 2009. Biogrout, ground improvement by microbial induced carbonate precipitation, 202 pp PhD thesis, Department of Biotechnology, Delft University of Technology, The Netherlands

75. Stocks-Fischer S., Galinat J. K., Bang S. S. 1999. Microbiological precipitation of $CaCO_3$. Soil Biol. Biochem. 31, 1563–1571 (doi:10.1016/S0038-0717(99)00082-6)10.1016/S0038-0717(99)00082-6

76. Karatas I., Kavazanjian E., Rittmann B. E. 2008. Microbially induced precipitation of calcite using *Pseudomonas denitrificans*. Proc. of 1st Int. Conf. on Bio-Geo-Civil Engineering, 23–25 June 2008, Delft, The Netherlands, pp. 58–66

77. van Paassen L. A., Daza C. M., Staal M., Sorokin D. Y., van der Zon W., van Loosdrecht Mark C. M. 2010. Potential soil reinforcement by biological denitrification.Ecol. Eng. 36, 168–175 (doi:10.1016/j.ecoleng.2009.03.026)10.1016/j.ecoleng.2009.03.026

78. van Paassen L. A., Ghose R., Van der Linden T. J. M., Van der Star W. R. L., Van Loosdrecht M. C. M. In press Quantifying bio-mediated ground improvement by ureolysis: a large scale biogrout experiment. ASCE JGGE. (doi:10.1061/(ASCE)GT.1943-5606.0000382)10.1061/(ASCE)GT.1943-5606.0000382

79. Warthmann R., Lith Y. v., Vasconcelos C., McKenzie J. A., Karpoff A. M. 2000.Bacterially induced dolomite precipitation in anoxic culture experiments. Geology 28, 1091–1094 (doi:10.1130/0091-7613(2000)28<1091:BIDPIA>2.0.CO;2)10.1130/0091-7613(2000)28<1091:BIDPIA>2.0.CO;2

80. Roden E. E., Leonardo M. R., Ferris F. G. 2002. Immobilization of strontium during iron biomineralization coupled to dissimilatory hydrous ferric oxide reduction. Geochim. Cosmochim. Acta 66, 2823–2839 (doi:10.1016/S0016-7037(02)00878-5)10.1016/S0016-7037(02)00878-5

81. Bachmeier K. L., Williams A. E., Warmington J. R., Bang S. S. 2002. Urease activity in microbiologically-induced calcite precipitation. J. Biotechnol. 93, 171–181 (doi:10.1016/S0168-1656(01)00393-5)10.1016/S0168-1656(01)00393-5

82. DeJong J. T., et al. 2009. Upscaling of bio-mediated soil

improvement. Proc. 17th Int. Conf. on Soil Mechanics and Geotechnical Engineering, 5–9 October 2009, Alexandria, Egypt, *pp. 2300–2303. Rotterdam, The Netherlands: Millpress Science Publishers.*

83. Mortensen B. M. 2008. Treatment optimization for microbially-induced calcite precipitation. Masters of Science, Civil and Environmental Engineering, University of California, Davis, CA

84. Barkouki T. H., Martinez B. C., Mortensen B. M., Weathers T. S., DeJong J. T., Spycher N. F., Ginn T. R., Smith R. W., Fujita Y. In press Forward and inverse bio-geochemical modeling of microbially induced calcite precipitation in 0.5 m columnar experiments. Transp. Porous Media.

85. Whiffin V. S., van Paassen L. A., Harkes M. P. 2007. Microbial carbonate precipitation as a soil improvement technique. Geomicrobiol. J. 25, 417–423

86. Lee J. S., Santamarina J. C. 2005. Bender elements: performance and signal interpretation. ASCE J. Geotech. Geoenviron. Eng. 131, 1063–1070 (doi:10.1061/(ASCE)1090-0241(2005)131:9(1063))10.1061/(ASCE)1090-0241(2005)131:9(1063)

87. Mitchell J. K., Soga K. 2005. Fundamentals of soil behavior. Hoboken: John Wiley & Sons Ltd

88. van Paassen L. A., Isimite J. O., Picioreanu C., van Loosdrecht M. C. M. 2008.Computational model to study microbial carbonate precipitation in porous media at micro-scale. Proc. 1st Int. Conf. on Bio-Geo-Civil Engineering, 23–25 June, The Netherlands, pp. 45–50, Delft, The Netherlands: TU Delft.

89. Xu T. 2008. Incorporation of aqueous reaction kinetics and biodegradation into TOUGHREACT: application of a multi-region model to hydrobiogeochemical transport of denitrification and sulfate reduction. Vad. Z. J. 7, 305–315 (doi:10.2136/vzj2006.0130)10.2136/vzj2006.0130

90. Xu T., Sonnenthal E. L., Spycher N., Pruess K. 2006. TOUGHREACT: a simulation program for non-isothermal multiphase reactive geochemical transport in variably saturated geologic media. Applications to geothermal injectivity and CO_2 geological

sequestration. Comput. Geosci. 32, 145–165 (doi:10.1016/j.cageo.2005.06.014)10.1016/j.cageo.2005.06.014

91. Xu T., Sonnenthal E., Spycher N., Pruess K. 2008. Toughreact users' guide: a simulation program for non-isothermal multiphase reactive geochemical transport in variably saturated geologic media. LBNL Report 55460-2008 Lawrence Berkeley National Laboratory, Berkeley, CA
92. Parkhurst D. L., Appelo C. A. J. 1999. User's guide to PHREEQC (Version 2)—a computer program for speciation, batch-reaction, one-dimensional transport, and inverse geochemical calculations, 310 p US Geological Survey Water-Resources Investigations Report 99-4259. Reston: VA; Department of the Interior & US Geological Survey
93. Fidaleo M., Lavecchia R. 2003. Kinetic study of enzymatic urea hydrolysis in the pH range 4–9. Chem. Biochem. Eng. 17, 311–318
94. Poeter E. P., Hill M. C., Banta E. R., Mehl S., Christensen S. 2005. UCODE_2005 and six other computer codes for universal sensitivity analysis, calibration, and uncertainty evaluation, techniques and methods. 6-A11, US Geological Survey, Reston, Virginia
95. Weil M. T., DeJong J. T., Mortensen B. M., Martinez B. C. Submitted Resistivity and sesimic measurements for real-time monitoring of microbially induced calcite precipitation in sands. ASTM Geotech. Test. J.
96. Abdel Aal G. Z., Atekwana E. A., Slater L. D., Atekwana E. A. 2004. Effects of microbial processes on electrolytic and interfacial electrical properties of unconsolidated sediments. Geophys. Res. Lett. 31, 4 p
97. Mortensen B. M., DeJong J. T., Martinez B. C., Weil M., Waller J., Gerhard R., Perderson L. Submitted Fabrication, operation, and health monitoring of bender elements for aggressive environments. ASTM Geotech. Test. J.
98. Battiato A., Tartakovsky D. M., Tartakovsky A. M., Scheibe T. 2009. On breakdown of macroscopic models of mixing-controlled heterogeneous reactions in porous media.Adv. Water Resour. 32,

1664–1673 (doi:10.1016/j.advwatres.2009.08.008)10.1016/j.advwatres.2009.08.008

99. Stumm W., Morgan J. J. 1996. Aquatic chemistry. 3rd edn New York, NY: Wiley-Interscience
100. Tartakovsky A., Redden G., Lichtner P., Scheibe T., Meakin P. 2008. Mixing-induced precipitation: experimental study and multi-scale numerical analysis. Water Resour. Res. 44, W06S04. (doi:10.1029/2006WR005725)10.1029/2006WR005725
101. IEA 2009. World energy outlook 2009. Austria: International Energy Agency
102. IPCC 2007. Fourth assessment report: climate change. Spain: Intergovernmental Panel on Climate Change
103. WCED 1987. Our Common Future. In Report of the World Commission on Environment and Development, World Commission on Environment and Development(WCED) (ed. Bruntland G., editor.). Oxford, UK: Oxford University Press
104. BERR 2008. Strategy for sustainable construction, business, enterprise & regulatory reform (BERR). Burton: HM Government
105. ICE 2009. The state of the nation, low carbon infrastructure. India: Institution of Civil Engineers
106. Glaser S. D., Shoureshi R., Pescovitz D. 2005. Future sensing systems. Smart Struct. Syst. 1, 103–120
107. Hoult N. A., Bennnet P. J., Stoianov I., Maksimovi C., Middleton C. R., Graham N. J. G., Soga K. 2009. Wireless sensor networks: creating 'smart infrastructure'. Proc. ICE, Civil Eng. 162, 136–143 (doi:10.1680/cien.2009.162.3.136)10.1680/cien.2009.162.3.136
108. Spencer B. F., Jr, Ruiz-Sandval M. E., Kurata N. 2004. Smart sensing technology: opportunities and challenges. Struct. Control Health Monit. 11, 349–368 (doi:10.1002/stc.48)10.1002/stc.48

Chapter 9

Transparent 2d/3d Half Bird's-Eye View of Ground Penetrating Radar Data Set in Archaeology and Cultural Heritage

Selma Kadioglu[1,2]

[1]Ankara University, Faculty of Engineering, Department of Geophysical Engineering, Ankara, Turkey

[2]Ankara University, Earth Sciences Application and Research Center, Ankara, Turkey

INTRODUCTION

This chapter focuses a new visualization approach to the monitoring of internal micro-discontinuities such as cracks, micro fractures and cavities, and archaeological remains with foundational infrastructure. The method uses a hybrid interactive two-dimensional / three-dimensional (2D/3D) transparent visualization of ground penetrating radar (GPR) data set gathered from sites of archaeological and cultural heritage. The data visualization is based on methodological formulation of amplitude–colour scale function for 2D radargram visualization to indicate micro-discontinuities, infrastructures and buried archaeological remains. The transparent 3D imaging combines to half bird's-eye view was constructed from a processed parallel-aligned 2D GPR profile data set by using an opaque approximation instead of linear opacity. The amplitude–colour scale is balanced by the amplitude range of the buried remains within a proposed depth range, and appointed an opaque coefficient in order to differentiate buried remains from others. Interactive visualizations are conducted of transparent 3D half bird's-eye view of GPR data volumes.

Archaeological and cultural heritage represent cultural identity and a source of creativity for present and future generations. Conservation of art treasures is a serious problem, since all objects change or deteriorate over time, mainly due to natural forces of decay. Requirements for maintaining the existing condition of buildings, monuments or statues of historical interest differ from the requirements of an initial treatment. Many of the problems associated with treatment involve the lack of prior, baseline information. Identification of the causes of degradation and understanding of the cause/effect relationships are crucial for the conservation of archaeological and cultural heritage. The treatment of such items generally involves coating or vapour barriers, and compatibility with substrate. However, cultural heritage requires maintenance not only for its walls, but also its infrastructures and security of foundations. Therefore, safety and maintenance management of such sites must include imaging of infrastructures and potential discontinuities. The same importance should be given to archaeological sites, especially in urban areas. Therefore, there is also a need for improved non-invasive methods of visualization in evaluating the progress of the buried infrastructures of archaeological heritage.

Ground penetrating radar (GPR), which is also called surface penetrating radar, is a time-dependent, high frequency electromagnetic geophysical technique that can provide a 3D pseudo image of the subsurface, including the fourth dimension of colour, and can also provide accurate depth estimates for many common subsurface objects [1, 2]. GPR uses the scattered wave field of high frequency electromagnetic (EM) waves. The EM waves travel at a specific velocity that is determined primarily by the permittivity of the material. The principles of GPR have been explained extensively in the literature [3], especially for fault and fracture imaging [4-9]; in assisting contaminated sites by locating buried features of interest such as underground storage tanks, pipes [10-11], unexploded ordnance (UXO) and clutter [12-14]; and in the mapping of shallow stratigraphy and discontinuities [3, 15-20].

Ground penetrating radar (GPR) provides more detailed results than other geophysical methods, because it can image the position and the depth of targets within very complex and restricted areas. The method is non-destructive and can be applied on a surface, a wall, or a monument [21-23]. The method can also be used in urban areas or in archaeological structures and, depending on the antennas and the particular situation, can achieve a resolution of the order of several centimetres 24-26]. Therefore, it has been the most commonly used method for defining cultural heritage and buried remains at archaeological sites. Furthermore, detailed imaging has become an important area of interest [1, 24- 28]. Generally, parallel 2D profile data are acquired in the archaeological site. 3D data imaging, obtained by aligning parallel 2D profile data sets, is used to identify temporal changes at a constant depth. The locations and the depth of the remains in the study area can be determined via slices of the 3D data volume. Therefore, the GPR method gives more precise results than other geophysical methods. However, the obtained results and their interpretation can be further improved when the data set is visualized as a volumetric rendering of the remains. This method allows anyone to imagine how an area looked by looking into the 3D image.

The aim of this chapter is to show transparent 3D GPR data visualization with the most suitable viewing angle into this 3D data volume including buried objects, which is called a transparent 3D half bird's-eye view of the GPR data volume or its sub-volumes. Therefore, first, we introduce the study areas and data acquisition, followed by general data processing steps of the gathered 2D GPR profile data set. Third, we show a revised colour range of the amplitude scale, representing the fourth dimension of the hybrid 2D/3D visualization. Fourth, we attempted to realize a new amplitude–colour-balancing approximation, according to the travel time range or depth range, as an alternative approximation of gain in order to prevent exaggerated amplitude gain, which affects transparency and obscures buried infrastructures. Fifth, we appointed a new opaque function, which must support amplitude–colour scale in order to supply transparency and reveal the fractures, cavities, buried wall remains and foundational infrastructures of interest. The fourth and fifth steps are conducted one within the other.

In this chapter, first, the method was used to show micro-discontinuities in monumental statue groups at Mustafa Kemal ATATÜRK's mausoleum (ANITKABIR) in Ankara, Turkey. Second, the method was used to define buried infrastructures and archaeological remains inside and northeast of the Zeynel Bey tomb, in Hasankeyf ancient city in south-eastern Turkey. The studies examined whether the proposed GPR method could yield useful results at these highly restricted sites. Our visualization results were an interpretation of the two datasets with different objectives.

MONITORING MICRO-DISCONTINUITIES IN MONUMENTAL STATUES

Study Area: Mustafa Kemal Ataturk's Mausoleum, Anitkabir

The Anitkabir mausoleum, was completed in 1953 in Ankara, Turkey (Figure 1). It hosts state ceremonies during national festivals, and represents the Turkish people and Ghazi Mustafa Kemal ATATÜRK, the founder of the Republic of Turkey (Figure 1b). Anitkabir was constructed in three phases. The first part is an entrance road, called the Lion Road, which is 262m long and has a total of 24 lion statues along each side, representing power and peace (Figure 1c). The second part is a ceremonial square, and the third is the Mausoleum (Figure 1c). At the beginning of the Lion Road, the Turkish people are represented by three large male statues in front of the Freedom Tower on the left side (Figure 1d), and three large female statues in front of the Independence Tower (Figure 1e) on the right side [29].

Figure 1: (a) Geographical location of the study area, (b) Ghazi Mustafa Kemal ATATURK, the founder of the Republic of Turkey, (c) the ANITKABIR monument, Ankara-Turkey, (d) three large male statues at the beginning of the Lion Road, which is the entrance to Anitkabir, on the left side, (e)female statues facing the male group, (f) 24 lion statues, representing power and peace, sit on each side of the Lion Road.

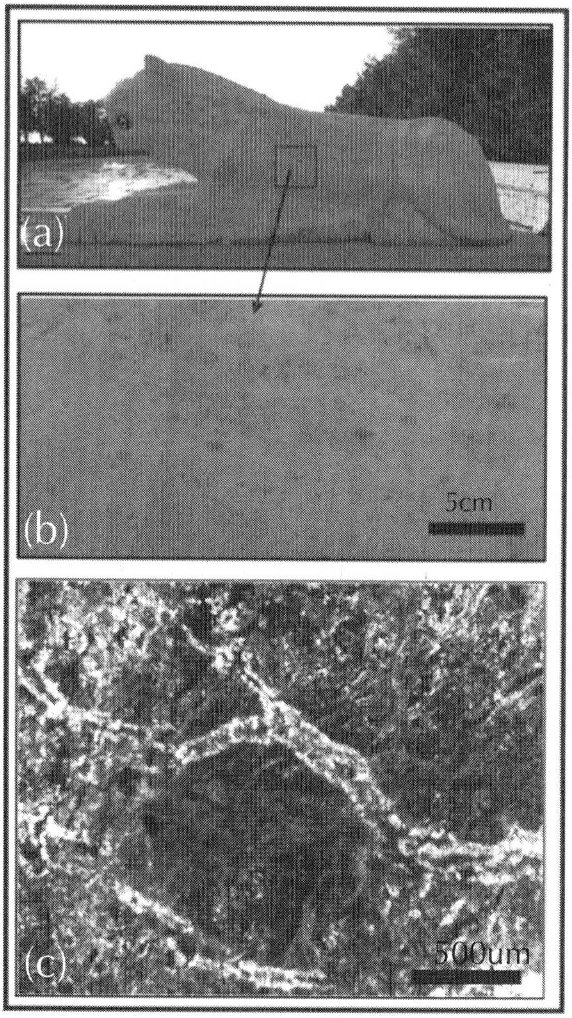

Figure 2: (a) A lion statue, (b) view close to surface of lion, and (c) microphotograph of white-coloured travertine.

The monument groups (three women, three men) and twenty-four lion statues of Anitkabir are mainly composed of white travertine from Pinarbasi, Kayseri, Turkey. The white-coloured travertine has a banded and spongy texture under the microscope [23]. It is mainly composed of calcite, aragonite with a small amount of salt, recrystallized calcite, gypsum and plant relicts. Table 1 shows the modal mineralogical composition and physical properties of this travertine. Micro-fractures were observed under a polarizing microscope, especially at the rim of the vesicular of the rocks (Figure 2).

Table 1: Mineralogical composition and physical properties of white-coloured travertine in Anitkabir

Mineral Composition	Calcite (54%) and Aragonite (31%)
Alteration Products	Recrystallized calcite (13.5) and rarely gypsum and halite (1.5%)
Colour	White
Hardness	3 Mohs, 52.8 Schmidt
Unit Volume Weight	2.52 gr/cm3
Porosity Ratio (%)	9.8 ± 2.180
Cracks Ration (%)	3.4
Alteration Ratio (%)	2.5

Figure 3: Data measurements on the female statues.

GPR Survey Description

The human statues were divided into several subparts such as skirt or under waist, between waist and neck, arms, trousers, etc. for GPR survey. Some parts were divided into additional subparts (Figures 3and 4) according to their figures, in order to enable true 3D imaging and protect the profile line position, because topography correction was not possible on the statues. In addition, three profiles spaced 10 cm apart were arranged along the backs of lion statues. The data acquisition scheme is shown in Figure 5, in which GPR data profiles gathered on the body of the first female statue were split into two parts, called skirt and upper part between waist and neck.

Profiles were spaced at 10 cm on each subpart, and were lined with a paper band sticker. First, data survey tests were carried out to determine the recording time-window according to the approximate thicknesses of the statues. A RAMAC CU II GPR system equipped with a 1.6-GHz bistatic shielded antenna was employed on all the statue groups. Transmitter–receiver antenna offset was 0.05m. Trace spacing was 0.0044m and time-sampling interval per trace on each profile was 0.0327 ns.

Figure 4: Data acquisitions on the male statues.

GPR Data Processing

Data processing was performed on 2D GPR profile data sets for each part of the statues with the REFLEXW program (ver. 5.5) developed by Sandmeier Scientific Software [30]. Start-time correction, then de-wow and background removals were applied to all the profile data in order to protect the true time scale, remove very low frequency effect and average amplitude knowledge respectively. The amplitude decay compensation was applied to all traces of the whole data set by using the same small-scale linear gain function. A second-order band-pass Butterworth filter was used for the whole data set to eliminate low-frequency artifacts and high-frequency noise. The resulting synthetic hyperbolas were matched with diffraction patterns throughout the profiles to determine average velocity of the electromagnetic (EM) wave. The best matching hyperbola provided the velocity of 0.12 m/ns. Finally, Kirchhoff migration was applied to the radargrams using average velocity, in order to carry diffracted electromagnetic (EM) waves true locations.

The quality of GPR images is strongly dependent on appropriate correction of the attenuation effects, usually supplied by time-varying gain. However, historically, the use of amplitude gain in basic processing of GPR data has been highly subjective and also very much displaying methodology [31]. There are various methods available for amplitude gain for GPR data. Traditional time-varying gain is carried out using linear, exponential functions, etc. functions, including ground wave amplitudes. However, this operation is not linear. The time-gained GPR data cannot recover the original information. Selection of the gain function depends mostly on the user and the quality of the GPR data. Both exaggerated linear and exponential time-gain change not only the amplitude range for each time step but also amplitude shape. Exaggerated time gain to image 2D GPR data can result in erroneous interpretation, especially when using such data to construct a 3D volume.

Figure 5: GPR data acquisition scheme of the first female statue.

Time-gained signal is

$$s(x,t) = r(x,t)h(t) \quad (1)$$

$$h(t) = a, 1 < a \leq |r\max r\min| \quad (2)$$

Where, h(t), is a time gain function, |rmax| and |rmin| are maximum- and minimum amplitude of the reflected/diffracted wave r(x,t), respectively. h(t), has a constant decimal value a between 1 and 2, and also is a linear function between 1 and a for a 2D data profile [32].

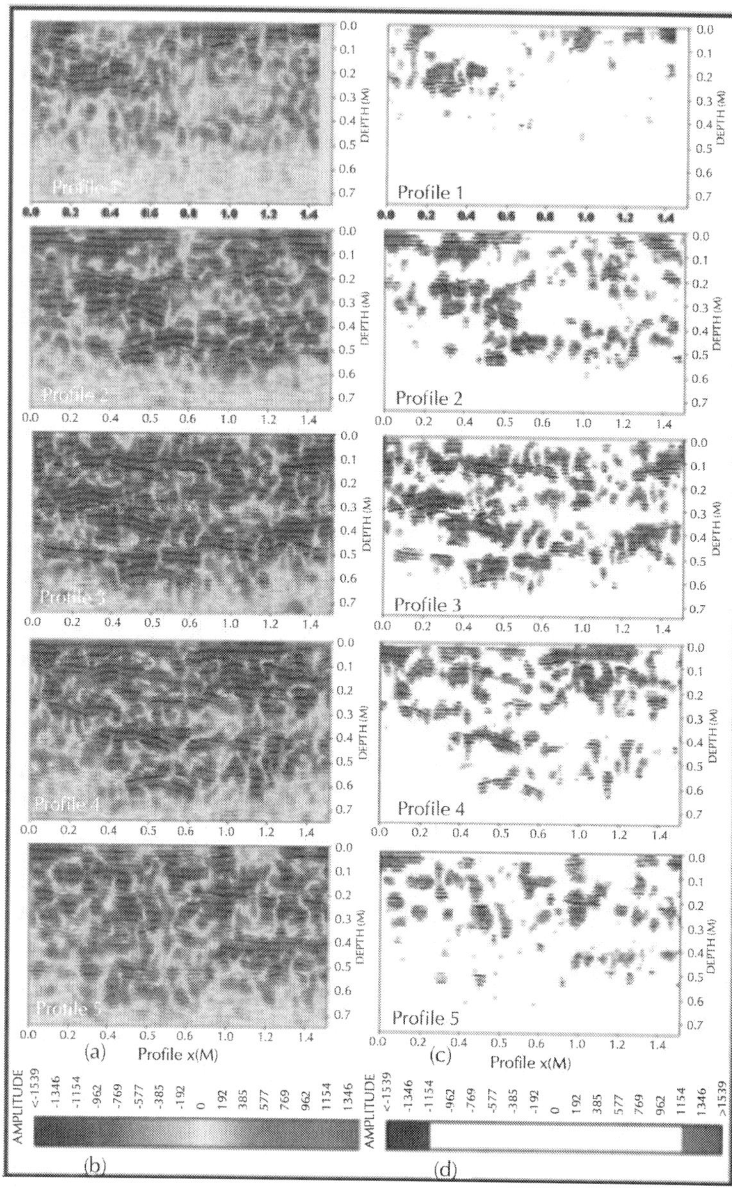

Figure 6: a) Processed 2D radargrams of the GPR profiles 1 to 5, acquired from the skirt of the first female statue, using (b) selected linear amplitude–colour function, (c) the same radargrams using (d) re-arranged amplitude–colour scale of (b) to reveal small cavities and fractures on the radargrams.

Our approximation concerns 2D or 3D image simplification in determining buried archaeological remains and discontinuities such as fractures. Therefore, we assigned a new amplitude–colour range for 2D radargrams of the GPR profiles and opaque range in order to identify anomalies and display the data set in a transparent 3D data volume. Figure 6a indicates some processed radargrams of the profiles of the skirt of the first female statue, shown in Figure 5. A linear colour scale of the radargrams (Figure 6b) indicates the amplitude range, which begins with maximum negative polarity and ends with maximum positive polarity. The blue colour range represents maximum negative amplitude, while the purple colour range represents the positive amplitudes according to their values. It is known that the maximum positive and maximum negative amplitude ranges in the amplitude–colour scale represent the fractures and cavities filled with air or buried archaeological remains in soil. Therefore, it is necessary to check the maximum amplitude ranges on the time slices to identify these target objects. We applied a new approximation to eliminate the weak reflections that are characteristic of cemented rock [7, 23], and to activate the fractures and cavities represented by strong reflections/diffractions on the radargrams (Figure 6c). This involved assigning a new colour scale for the amplitude range of the processed profile data set by means of a new amplitude–colour function rather than linear amplitude–colour function (Figure 6d, Figure 7).

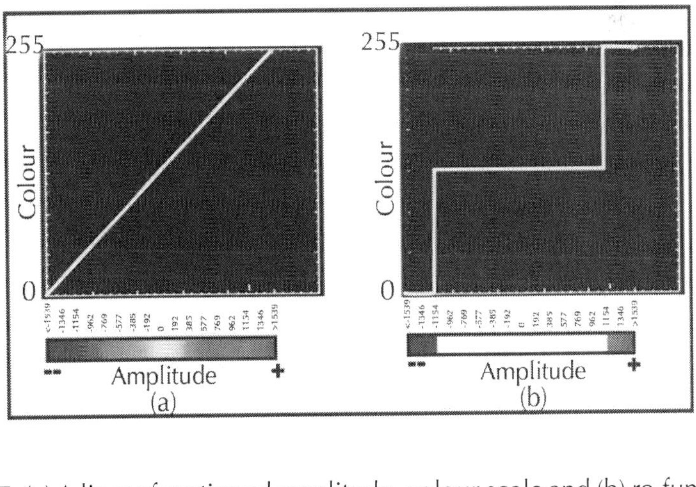

Figure 7: (a) A linear functioned amplitude–colour scale and (b) re-functioned colour scale of the same amplitude range.

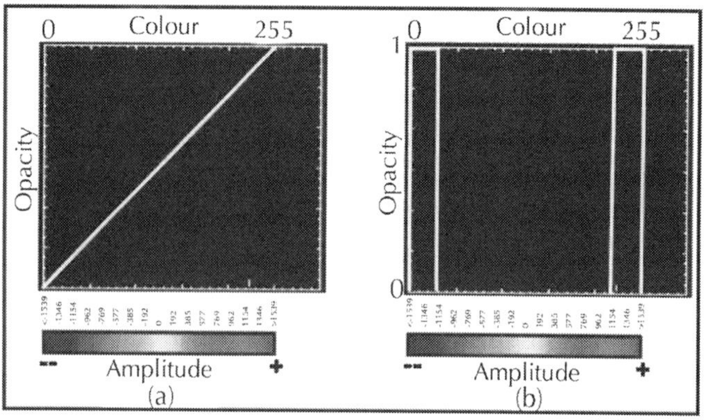

Figure 8: (a) A linear functioned opacity for the linear functioned amplitude–colour scale and (b) A re-arranged opacity function using the same amplitude–colour scale to activate only maximum amplitude range and remove all others.

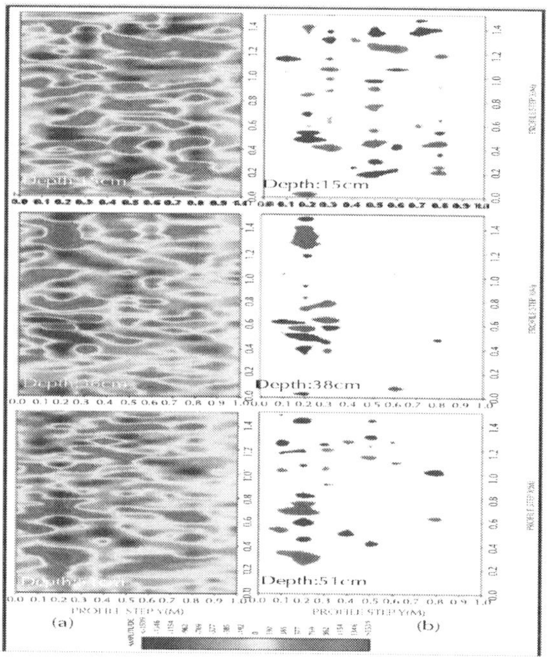

Figure 9: (a) Traditional solid depth slices at 15 cm, 38 cm and 51 cm, (b) Transparent depth slices for the skirt of the first female statue (same three depths).

The horizontal axis of the amplitude–colour function (Figure 7) is the amplitude scale of the GPR data, whereas the vertical axis represents colour categories from 0 to 255. The colour limitation allowed simplification and made only fractures and native cavities visible on the radargrams, as seen inFigure 6c. The horizontal x-axis of Figures 6a and 6c indicate the distance along the measuring profile. The vertical axis shows depth range, which represents the thickness of the statue skirt from front surface to back surface, transformed by using average velocity of the EM wave.

Transparent 3D Half Bird's-Eye View Of GPR Data Set

Generally, interactive visualization is carried out by constructing 3D data volumes of parallel-aligned 2D GPR data sets to show the target objects. The 3D data volume can be displayed as slices, including profiles, times (or depths) and common traces of the profiles; or separated sub-blocks are rendered as solid iso-volumes with linear opacity, determined by the amplitude of the anomalies. The buried fractures or cavities can be defined on the interactive slices, particularly on depth slices with location, and shapes according to depth. Therefore, it was necessary to check the most meaningful depth slices and profiles to define the structures of the subsurface if the area is small and complex. However, the obtained results could be further improved.

Our aim was to obtain a good 3D data volume display, which was a critical part of interpreting the GPR data set. The 3D image is able to present a view of subsurface features such as a fracture or cavity, in addition to objects such as industrial and/or archaeological remains, etc. This imaging could be achieved by a transparent 3D half bird's-eye view revealing only buried objects. Therefore, firstly, transparency could be achieved by constructing an opacity function instead of linear opacity determined by the amplitude scale (Figure 8). The horizontal axis of the opacity function was the amplitude scale starting with maximum negative amplitude and ending with maximum positive amplitude; the vertical axis represented opacity coefficients of the amplitude range [11, 23]. Thus, any amplitude range could be highlighted or minimized by the appointed opacity coefficient. The REFLEXW program allows the opacity coefficient to be chosen between one (maximum opacity) and zero (transparent) (Figure 8b) [26]. A transparent view could be obtained only by eliminating the unwanted amplitude range.

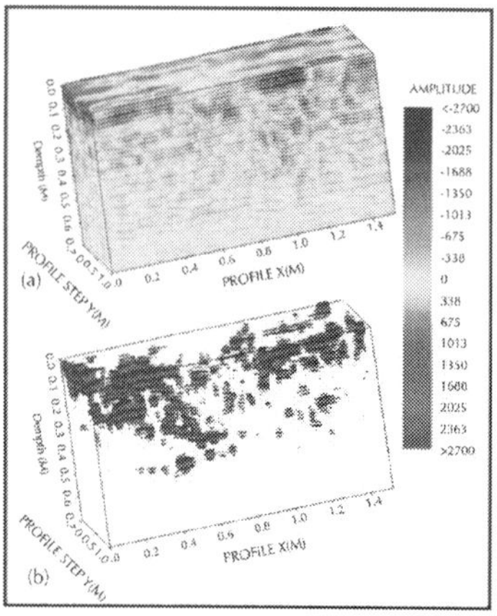

Figure 10: (a) Solid 3D GPR data volume visualization of the skirt of the first female statue (Figure 5) with all profiles and depth range through special view angle of profile slices, (b) transparent half bird's-eye views of the same 3D data volume of (a).

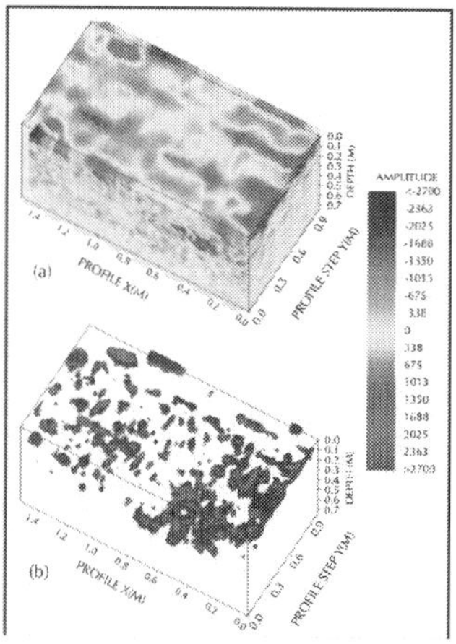

Figure 11: (a) Solid 3D data volume visualization of the skirt of the first female statue (Figure 5) with all profiles and depth range through special viewing angle of depth slices, (b) transparent half bird's-eye visualization of the same 3D data volume in (a).

Therefore, the amplitude range was important. Because it was known that the maximum amplitudes represented discontinuities, the weak amplitude range was eliminated by giving these a zero opacity value, and transparent 3D imaging was obtained. The transparency was achieved by allocating an opaque interval to the amplitude scale, similar to the re-arranged amplitude–colour approximation for interested profile range or time range for the solid 3D GPR data volume. This visualization type was applied to both the statues and archaeological remains in this chapter. Figure 9a indicates traditional, solid depth-slices, while Figure 9b indicates transparent depth slices at 15 cm, 38 cm and 51 cm of the data set for the skirt of the first female statue. These slices were used to control micro-fractures and cavities according to the skirt thickness. The horizontal x-axis and y-axis of slices indicate the profile sequence and the distance along the measuring profile respectively. Locations of the micro-cavities could be seen on the slices. It is necessary to carefully check interactive slices in order to determine location and shapes according to depth (thickness).

Secondly, it was necessary to arrange viewing angles of the x, y and z axes, to obtain the maximum meaningful 3D data volume for the relevant depth range or profile range. The slices could be interpreted differently with differing viewing angles, although the visualization of the slices was required to be the same with the data measurement axes on the map or on a picture. However, there was a general lack of knowledge about the subsurface, including the remains, and about when to take data measurements with regard to information such as the direction of a fracture or an archaeological wall. Therefore, the data measurement strategy was decided according to the field size.

The slices obtained with the same axes on the map of the study site could not effectively represent the subsurface. In addition, when the slices were rotated around the axes, a lining fracture or a wall along the same direction as the angle of view of the slice could be imaged more effectively than in the standard view. To image the fractures and native cavities in the statues, it was decided to use a transparent 3D sub-volume of the profile and depth slices with a half bird's-eye view by arranging the view angles of the axes. Figures 10a and 11a indicate the solid 3D data volume visualization of the skirt of the first female statue with all profiles and depth range using a special viewing angle of profiles and depth slices. In addition, Figures 10b and 11b represent our approximation with the transparent 3D half bird's-eye view visualization of the same data set. The horizontal x-axis of Figure 10 and Figure 11 indicates the distance along the profile. The horizontal y-axis represents the profile sequence. The vertical axis indicates thicknesses of statues from the front surface to the back surface of the skirt of the statue. Figure 12 shows different depth ranges of transparent 3D half bird's-eye views of the GPR data aligned on the first female statue (Figures 5 and 11b) between 0–10 cm, 10–20 cm, 20–30 cm 30–40 cm, 40–50 cm and 50–60 cm depth ranges, and shows the locations of the micro-fractures and cavities with purple and blue colours, represent the maximum amplitude ranges. Figure 13 indicates different profile ranges of transparent 3D half bird's-eye views of the GPR data aligned on the skirt of the first statue (Figures 5 and 10b) between profiles 1–3, profiles 4–6, profiles 7–9 and profiles 10–11; and the upper part of the female statue between profiles 1–3 and profiles 4–8 through special viewing angle of the profile slices; the locations of micro-fractures and cavities are represented by purple and blue colours.

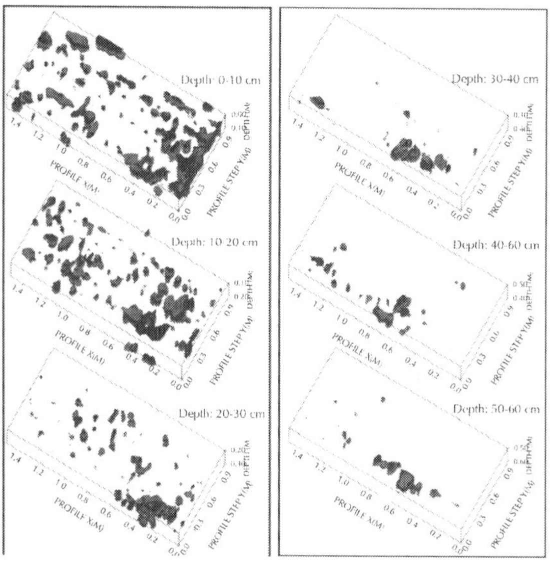

Figure 12: Transparent half bird's-eye view results of the 3D sub-volumes of Figure 11b between 0–10 cm, 10–20 cm, 20–30 cm, 30–40 cm, 40–50 cm and 50–60 cm depth ranges, respectively, including internal micro-fractures and native cavities.

Figure 13: Transparent half bird's-eye results of the 3D sub-volumes shown in Fig. 10b between profiles 1–3, profiles 4–6, profiles 7–9 and profiles 10–11; and the upper part between profiles 1–3 and profiles 4–8 through special viewing angle of profile slices, including internal micro-fractures and cavities.

The profile ranges of Figure 13 give the locations of the fractures and cavities throughout the depth of the profile ranges, while the depth ranges of Figure 12 give the locations of the fractures and cavities along the full surface of the skirt through the depth ranges. Therefore, it is possible to check both transparent profile ranges and depth ranges according to the most appropriate viewing angle of the 3D GPR data volume in order to determine the locations and continuities of micro-fractures and cavities within very restricted study areas, such as those in the present study.

The native micro-cavities are not effective to harm the lions. The micro-fractures show a lateral, inclined or vertical linearity. According to Figures 12 and 13, the skirt of the statue had an important fracture between profiles 1 and 3, ranging from 0–80 cm from the front surface as far as the back surface. In addition, the figure had more small fractures and native cavities between 0- and 30-cm depth. To summarize to our method, we present results from the transparent 3D half bird's-eye view of three GPR data sets gathered form the backs of three lion sculptures, using three parallel-aligned profiles along the leg and head to visualize interior fractures (Figure 14).

According to the visualization results in the first lion, there were three large fractures aligned parallel to the surface along the back legs, which continue from the upper surface of the back legs to the border lion along the depth; and there were native cavities in the back through the belly. The second lion mostly had disorderly native cavities and micro-fractures along the back surface until the belly. The last lion was seen very powerful, and only included some native cavities in the back side.

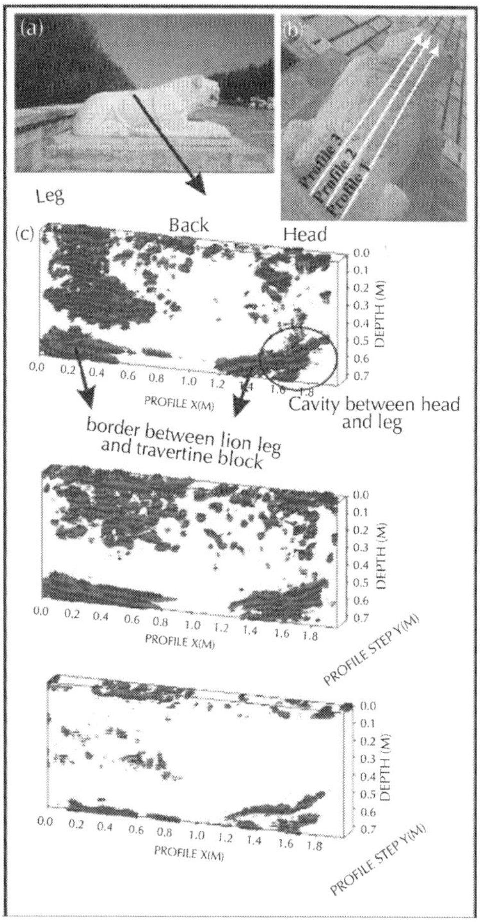

Figure 14: (a) One of the 24 lions on the Lion Road, (b) data acquired from three parallel profiles from the back to the head of the lion, (c) the results of the transparent half bird's-eye view of the three different lions.

PICTURING BURIED ARCHAEOLOGICAL REMAINS AND FOUNDATIONAL INFRASTRUCTURES

Zeynel Bey Tomb in Hasankeyf Ancient City

Our new 3D visualization was applied to archaeological remains both inside and outside the Zeynel Bey tomb in the ancient Turkish city of Hasankeyf. This site is among the last remaining locations of the Silk Road in Anatolia, spreading towards the East, and is located in Batman province, southeastern Turkey (Figure 15). A similar type of visualization for archaeological remains was introduced by previous studies [26, 33].

The precise foundation date of Hasankeyf is not known. The geopolitical situation in Hasankeyf strengthens the possibility of its being a very ancient settlement area. Hasankeyf is identified with the tomb built by Uzun Hasan for his son Zeynel Bey, who died in the war of Otlukbeli (1473) by the Tigris [34]. The Zeynel Bey tomb, the first example of the Anatolian mausoleum tradition (Figure 16), is on the north bank of the Tigris, across from the city.

The tomb is a cylinder of diagonal patterns made using brick and tile, with a pointed arch portal doorway on the north and a window in the south wall (Figure 16). Above the main shaft is a slightly smaller diameter shaft, which has small windows in each of the cardinal directions and carries a hemispherical dome (Figures 16 and 17) [34-37]. Inside, the plan is octagonal, with muqarnas niches supporting the transition to the round base of the dome. Each of the eight walls has a rectangular arched niche, and the burial chamber is recessed into the floor (Figure 17) [36].

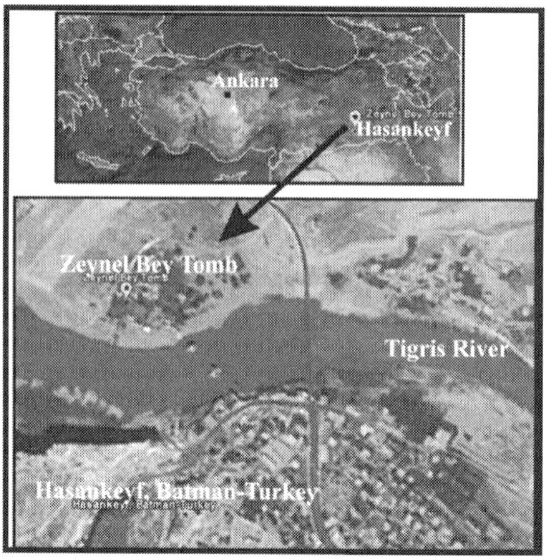

Figure 15: Geographical map of the Zeynel Bey tomb in Hasankeyf ancient city, Turkey.

Figure 16: Appearance of the Zeynel Bey Tomb, the first example of Anatolian mausoleum tradition, with some architectural details.

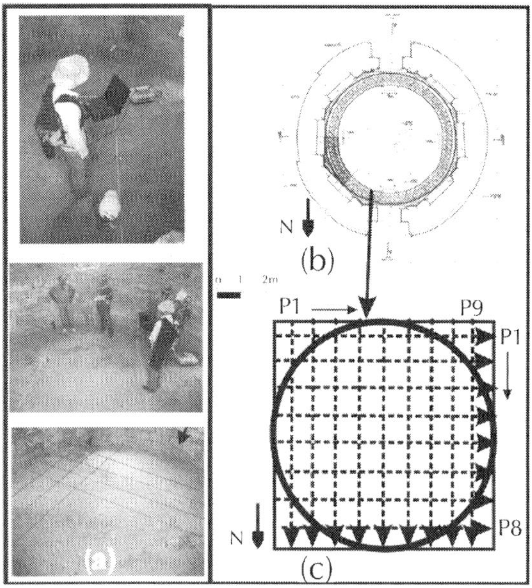

Figure 17: (a) GPR data measurements inside the Zeynel Bey Tomb, (b) Interior plan of Zeynel Bey Tomb: octagonal, with muqarnas niches supporting the transition to the round base of the dome, (c) The data measurement plan inside the tomb.

Hasankeyf and many other Tigris Valley settlements that have previously directed world history will be submerged when the proposed Ilisu Dam is completed. Despite proposals to move the monuments, many historic sites and artifacts will be lost when the reservoir is filled. Efforts to relocate and preserve culturally significant sites are currently led by Prof. Dr. Abdüsselam Uluçam, the rector of Batman University [37].

GPR Data Measurements at the Zeynel Bey Tomb

In this section, we present only two parts of the huge study area, including the tomb of Zeynel Bey and the Ottoman bath, an area of approximately 150×200m. The first part was inside of the tomb. The Zeynel Bey tomb is 4m along the east–west orientation and 3.5 m along the south–west (Figure 17). A RAMAC CUII GPR system was used with a bi-static 500-MHz center band shielded antenna to acquire the profile data. Within the tomb, 9 parallel profiles spaced 0.5m apart were directed from east to west, and 8 profiles were directed from north to south, making a total of 17 profiles (Figure 17c). The second part of the survey was conducted on the northeast side of the tomb, and 19 profiles were directed from south to north on the east side (Figure 18). Parallel profiles were spaced 1m apart, and each profile had a trace spacing of 5cm and a 70ns time-window per trace.

Data Processing and a New Amplitude-Balancing Approximation for Transparent 3d Half Bird's-Eye View of the GPR Data Set

The GPR data, gathered within and on the northeast side of the tomb, were processed using REFLEXW software. After sequencing the profiles as discussed at the Anitkabir site, the start-time correction was applied. De-wow and background removals were applied. The second-order band-pass Butterworth filter was then applied to the whole data set, to eliminate unwanted frequency noise. A simple linear gain function was applied as discussed in section 2.3. Velocity analysis indicated that the average velocity of electromagnetic wave propagation was 0.11m/ns. Finally, Kirchhoff migration was applied to the data.

Figure 18: (a) GPR data measurement on the northeast side of the Zeynel Bey tomb.

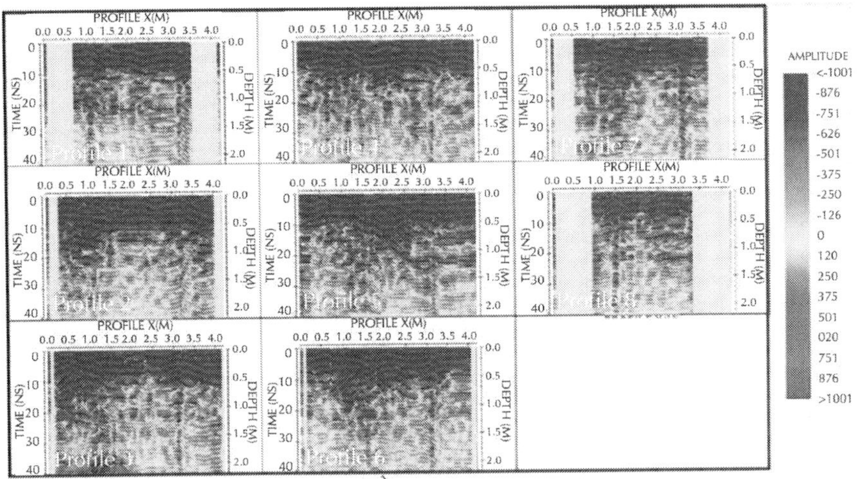

a)

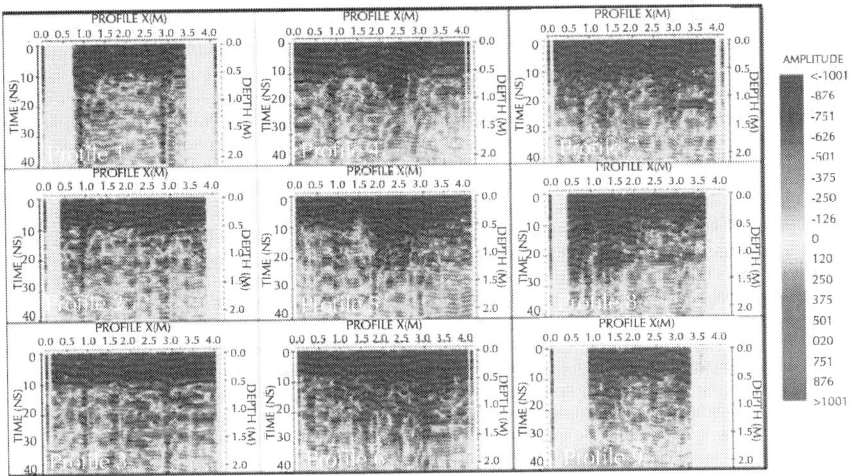

b)

Figure 19: Processed radargrams of the profiles along (a) the east–west direction and (b) the south–north direction inside the Zeynel Bey tomb.

The remains of a buried archaeological wall and foundational infrastructures can be defined on depth slices with location, and shapes according to depth. However, depth slices may not explain the subsurface if the area is small and complex, as in the Zeynel Bey tomb. Therefore, it is necessary to check the most meaningful depth slices and profiles to define the subsurface structures. During the control of the interactive depth slices, it is also important to control the colour of the amplitude range of the remains. It is known that the maximum amplitude ranges on an amplitude–colour scale represent subsurface features. However, the amplitude decreases with increasing recording time even if an appropriate time gain is applied to the profile data.

It is also known that if the colours of maximum amplitude ranges are dominant at the beginning of the recording time or at depths very near to the surface, then the resolution of the slices can not be sufficiently high to differentiate buried structures very near to the surface, especially for small study areas as seen on Figure 19. Our third approximations after re-arranged opacity function and viewing angle into the 3D volume, related to achieving amplitude balance according to the recording time or the depth (Figure 20). The aim of the amplitude balancing was to protect the colour range of the anomaly representing the target remains according to the time or depth range of the 3D sub-volumes; this was achieved by increasing or decreasing the maximum amplitude values of the amplitude–colour scale for the special depth slice or for a depth range. Therefore, the full data volume should be divided into sub-data volumes in time or depth.

To balance the amplitude–colour scale in a depth range, firstly the changing colour range of the remains was controlled. Then, the amplitude range of the remains was determined in the 3D sub-volume and the maximum amplitude values of the colour range were reduced or increased in the depth range. The balancing of the amplitude–colour scale according to the amplitude range of the buried wall remnants produced better resolution and showed the remains with the same colour ranges on the slices or their 3D sub-volumes with increasing time or depth axis. The balancing procedure was also a time gain, which was a non-uniform stair function, weighting the electromagnetic wave field according to the time axis (Figure 20). The approximation was important to obtain a 3D representation of the volume.

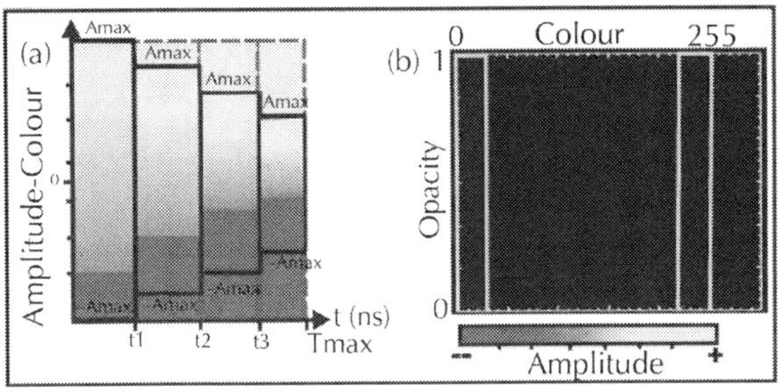

Figure 20: (a) Re-scaling maximum amplitudes according to selected time or depth range of 3D GPR sub-volumes, and assigning the same colour range; (b) assigning a new opaque range according to the re-scaled amplitude–colour range.

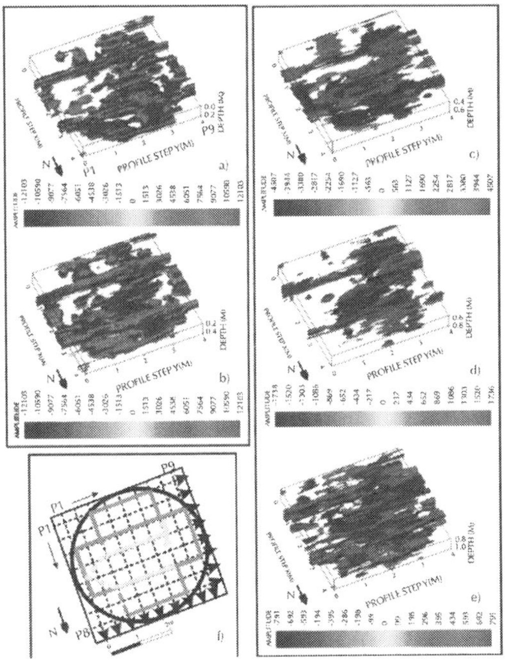

Figure 21: Transparent 3D half bird's-eye views of the GPR data set aligned in the south–north direction inside the Zeynel Bey Tomb between (a) 0 and 20 cm, (b) 20 and 40 cm, (c) 40 and 60 cm, (d) 60 and 80 cm, and (c) 80 and 100 cm depth ranges; and (f) infrastructures (orange colour) of the base and a cemetery (yellow colour) in the base wall inside the tomb.

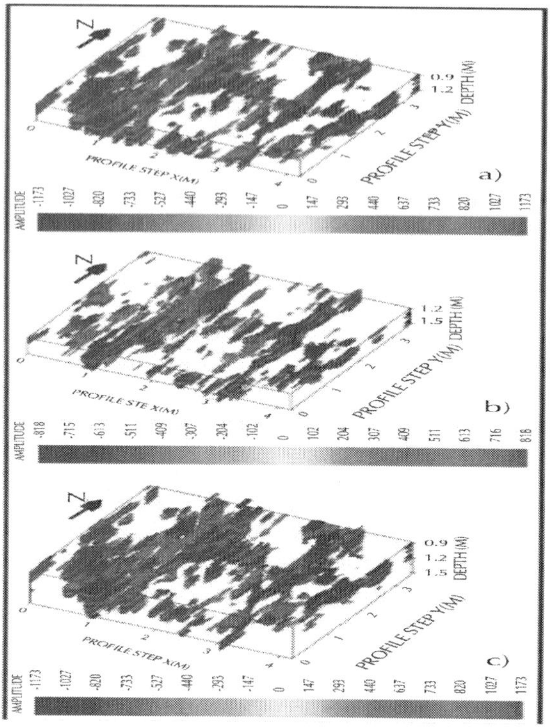

Figure 22: Transparent 3D half bird's-eye views of the GPR data set aligned east–west inside the Zeynel Bey tomb at depths between (a) 90 and 120 cm, (b) 120 and 150 cm, and (c) 90 and 150 cm.

Practically, it was only possible to highlight the amplitude–colour ranges representing the remains, which were previously established as maximum amplitude–colour ranges of the 3D sub-data volumes. It was therefore straightforward to construct an opacity function by appointing opacity coefficients of one (maximum opacity) for maximum negative and maximum positive amplitude–colour ranges and zero (transparent) for unwanted amplitude–colour ranges on the amplitude–colour scale. A transparent 3D half bird's-eye view was obtained only by eliminating unwanted amplitude range to reveal subsurface remnants and features. However, depth range depended on the maximum amplitude volume. In order to construct a large depth range, the maximum amplitude–colour range would need to be restricted more than the normal range to produce a meaningful image. Figure 21 shows transparent 3D sub-data volumes with half bird's-eye view between 0 and 20cm, 20 and

40 cm, 60 and 80 cm, and 80 and 100cm depth ranges from the GPR data set aligned in the south–north direction inside the tomb. Wall structures and a cemetery with stair could be clearly seen. In addition, Figure 22shows transparent 3D sub-data volumes between 90–120 cm. 120–150 cm and 90–150cm depth ranges along the east–west direction inside the tomb. The visualization results of the east–west side data set supported those of the south–north. However, the results of the south–west side perfectly imaged the foundational infrastructure and buried cemetery.

Figure 23: (a) The North side of the Zeynel Bey tomb, (b) the excavation on (a) and a cemetery, (c) the walls of the cemetery and stairway.

Although the maximum diameter inside the tomb was 4 m, it would be possible for users to envisage buried archaeological remains when viewed using the transparent 3D image. Figure 23 shows the excavation results inside the tomb. The excavation findings perfectly matched the transparent 3D half bird's-eye view. The visualization results also precisely indicated all details of the foundational wall remains and the cemetery with stairs.

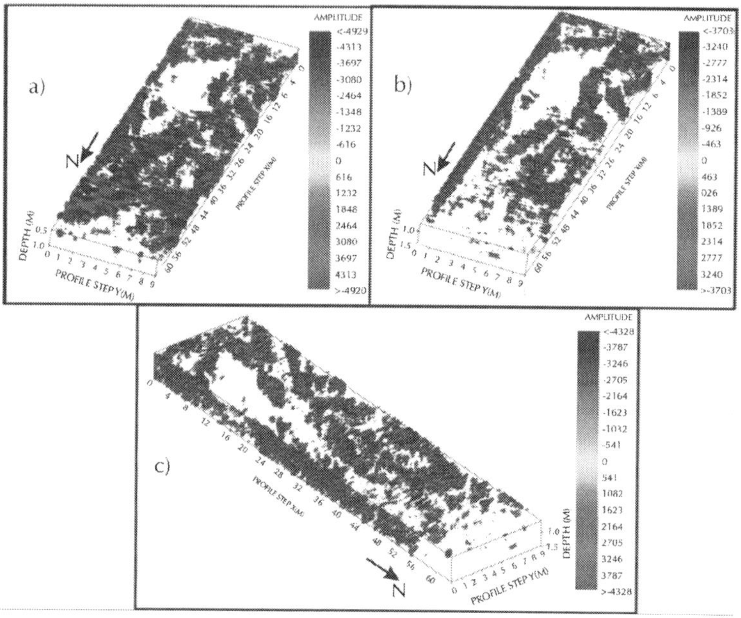

Figure 24: Transparent 3D half bird's-eye views of the GPR data set on the northeast side of the Zeynel Bey tomb, and pictured buried wall remains in depth between (a) 50 and 100 cm, (b) 75 and 150 cm; (c)75 and 150 cm but with different viewing angles of the x, y and z axes.

Half bird's-eye views of the transparent 3D depth volume ranges of the GPR data were also produced for the northeast side of the tomb (Figure 24). The wall structures were seen exactly in the corresponding transparent 3D imaging. The results showed that the buried walls were very near the surface and very complex. These imaging results remain to be confirmed through direct observation, as excavation has not yet taken place at these sites. The results could be confirmed by comparing the excavations inside of the tomb and their 3D visualization results.

RESULTS AND CONCLUSIONS

Interactive transparent 3D visualizations of GPR data volumes were produced for each part of 6 human statues and a group of 24 lion sculptures at Anitkabir, Turkey. We produced images of natural internal cavities and fractures within the statues via a proposed method of interactive transparent 3D half bird's-eye visualization of the GPR data volume of the profile range and the depth range.

Our first approximation related to amplitude–colour simplification. This allowed us to differentiate and locate native micro-cavities and micro-fractures inside the statues, based on radargrams of the profiles gathered on the surface of the statues. The second approximation related to our proposed opacity function, which dominated maximum positive and negative amplitudes and eliminated other irrelevant amplitudes. Third, an interactive transparent 3D half bird's-eye view of the 2D GPR data set was achieved by carefully assigning the amplitude–colour scale and its opacity range, together with a carefully selected viewing angle, profile and/or depth range. The transparent 3D imaging proved successful in identifying changes in the statues, accurate *x–y* locations and accurate depths. This monitoring replied to the aim of the study. Mapping fractures and cavities within statue groups could enable evaluation of their stability and indicate the best way to minimize restoration costs.

In the fourth approximation, the present study developed an improved amplitude-balancing approximation in order to reveal and differentiate subsurface historical remains from the surrounding soil medium. By combining the second and third approximations, the balancing of the amplitude–colour scale achieved sufficiently high resolution to represent the remains with the same colour ranges on both the slices and their transparent 3D imaging with increasing depth axis. A viewing angle was allocated to the x, y and z axes of the 3D data-volume by checking wall orientations and the data measurement directions on the map to obtain a good half bird's-eye view. The transparent 3D half bird's-eye view of the 2D GPR data set provided better imaging to accurately visualize the subsurface by sensing x–y locations and depths. The results demonstrated that the GPR method and our developed 3D visualization gave perfect results in a closed, circular area including remnants of very complex buried foundational wall, including a cemetery, within the Zeynel Bey tomb without any risk of damage to the tomb and infrastructures. In addition, the results also indicated that the visualization method perfectly monitored the archaeological buried walls with accurate x–y locations and accurate depths.

We also indicated that GPR method provides highly accurate results for the position and depth of targets within very complex and restricted areas, even on statues.

ACKNOWLEDGEMENTS

The authors would like to thank the military authorities at ANITKABIR; KA.BA Conservation of Historic Buildings and Architecture Ltd. with the coordination of rescue excavation director Prof. Dr. Abdüsselam Uluçam, Rector of Batman University, Turkey; and Ankara University Earth Sciences Application and Research Center (YEBIM) for supporting the projects. The authors also thank Prof. Yusuf Kagan Kadioglu, head of the Geological Engineering Dept., Ankara University, Turkey; Asst. Prof. Ali Akin Akyol, Baskent Vocational Higher School Program of Restoration and Conservation, Ankara University; and the MSc student group of the Ankara University Geophysical and Geological Engineering Departments for helping us in data collection at Anitkabir.

REFERENCES

1. Daniels JJ. Ground Penetrating Radar for Imaging Archaeological Objects in the Subsurface: Proceedings of the New Millennium International Forum on Consideration of Cultural Property, Kongju, Korea. 2000; 247-265.
2. Daniels DJ. Ground Penetrating Radar. The Institution of Electrical Engineers. Second Edition, London, United Kingdom, 2004.
3. Davis JL, Annan AP. Ground-Penetrating Radar for High Resolution Mapping of Soil and Rock Stratigraphy. Geophysical Prospecting 1989; 37 531–555.
4. Grandjean G, Gourry JC. GPR Data Processing for 3D Fracture Mapping in a Marble Quarry (Thassos, Greece). Journal of Applied Geophysics 1999; 36 19–30.

5. Orlando L. Ground Penetrating Radar in Massive Rock: A Case History. European Journal of Environmental and Engineering Geophysics 2002; 7 265–279.

6. Grasmueck M. 3-D Ground Penetrating Radar Applied to Fracture Imaging in Gneiss. Geophysics 1996; 61 (4) 1050–1064.

7. Grasmueck M, Weger R, and Horstmeyer H. Full-Resolution 3D GPR Imaging. Geophysics 2005; 70 (1) K12-K19.

8. Tsoflias GP, Gestel J-PV, Stoffa PL, Blankenship DD, Sen M. Vertical Fracture Detection By Exploiting the Polarization Properties of Ground-Penetrating Radar Signals. Geophysics 2004; 69 (3) 803–810.

9. Kadioglu S. Photographing Layer Thicknesses and Discontinuities in a Marble Quarry with 3D GPR Visualisation, Journal of Applied Geophysics 2008; 64 109-114.

10. Porsani JL and Sauck WA. GPR Profiles Over Multiple Steel Tanks:Artifact Removal Through Effective Data Processing. Geophysics 2007; 72(6) J77-J83.

11. Kadioglu S and Daniels JJ. 3D Visualization of Integrated Ground Penetrating Radar Data and EM-61 Data to Determine Buried Objects and Their Characteristics. Journal of Geophysics and Engineering 2008; 5 448-456.

12. Khadr N, Barrow BJ and Bell TH. Target Shape Classification Using Electromagnetic nduction Sensor Data *Proc. UXO Forum'98* Online at http://citeseer.ist.psu.edu/khadr98target.html

13. Pasapane BE and Sieling DR. Utilizing A Multiple Metal Detector Array for Locating Anomalies Inductive of Ferrous and Non-Ferrous OE/UXO 2002; 02P-0157, 28th Environmental &Energy Symposium & Exhibition (Ray F. Weston Inc., Charleston, South Carolina); 2002.

14. Won J, Keiswetter D and Bell T. Electromagnetic Induction Spectroscopy for Clearing Landmines. IEEE Transactions On Geosciences And Remote Sensing 2001; 39 703- 709.

15. Benson A K Applications of Ground Penetrating Radar In Assessing Some Geological Hazards: Examples of Groundwater Contaminants, Faults, Cavities J. Appl. Geophys. 1995; 33 177–93.

16. Yoder RE, Freeland RS, Ammons JT and Leonard LL. Mapping Agricultural Fields with GPR and EMI to Identify Offsite Movement of Agrochemicals. *J. Appl. Geophys* 2001; 47 251–9.
17. Stroh JC, Archer S, Doolittle JA and Wilding L. Detection of Edaphic Discontinuities with Ground-Penetrating Radar and Electromagnetic Induction. *Landscape Ecol.* 2001; 16 377–90.
18. Koralay T, Kadioglu S And Kadioglu YK. A New Approximation in Determination of Zonation Boundaries of Ignimbrite by Ground Penetrating Radar: Kayseri, Central Anotalia, Turkey. Environ. Geol. 2007; 52 1387–97.
19. Kofman, L, Ronen, A, Frydman, S. Detection of Model Voids by Identifying Reverberation Phenomena in GPR Records. Journal of Applied Geophysics 2006; 59 284–299.
20. Orlando L and Slob E. Using Multicomponent GPR to Monitor Cracks in a Historical Building J. Appl. Geophys. 2009; 67 327–34.
21. Masini N, Persico A, Guide A and Pagliuca A. A Multifrequency and Multisensory Approach for the Study and the Restoration of Monuments: The Case of the Cathedral of Matera. Advances in Geosciences 2008; 19 17–22.
22. Bavusi M, Di Napoli R and Soldovieri F. Microwave Tomographic Approach for Masonry Investigation: Some Real Results. Advances in Geoscience 2010; 24, 83–8.
23. Kadioglu S and Kadioglu, YK. Picturing Internal Fractures of Historical Statues Using Ground Penetrating Radar Method. Advances in Geosciences 2010; 24 23-34, www.adv-geosci.net/24/23/2010/.
24. Leucci G and Negri S Use of Ground Penetrating Radar To Map Subsurface Archaeological Features n An Urban Area. J. Archaeological Sci. 2006; 33, 502–12.
25. Masini N, Nuzzo L and Rizzo E. GPR Investigations for the Study and the Restoration of the Rose Window of Troia Cathedral (Southern Italy) Near Surf. Geophys. 2007; 5 287–300.
26. Kadioglu S. Definition of Buried Archaeological Remains with a New 3D Visualization Technique of Ground Penetrating Radar Data Set in Temple Augustus in Ankara, Turkey. Near Surf. Geophys. (Special issue on GPR in Archaeology) 2010; 8 397–406.

27. Goodman D Ground penetrating radar simulation in engineering and archaeology. Geophysics 1994; 59 224–232.

28. Goodman D, Schneider K, Piro S, Nishmura Y and Pantel AG. Ground penetrating radar advances in subsurfaces imaging for archaeology. In: Remote Sensing in Archaeology (eds J. Wiseman and F. El-Baz), 2007; 367–386, Springer.

29. http://www.tsk.tr/eng/anitkabir: website of the Turkish General Staff (TSK).

30. Reflexw. Sandmeier Scientific Software 2007

31. Annan AP, Practical processing of GPR data. Proceedings of Second Government Workshop on Ground Renetrating Radar, Sensor and Software Inc. 1999.

32. Kadıo lu S, Daniels JJ. Different Time Gain and Amplitude-Color Arranging for Ground Penetrating Radar Data: Applied Samples: XIII International Conference on Ground Penetrating Radar Lecce-ITALY, 2010. IEEE Xplore Digital Library Conference Puplications, Ground Penetrating Radar (GPR), 2010; 13th International Conference, E-ISBN : 978-1-4244-4605-6; DOI: 10.1109/ICGPR.2010.5550165, 2010.

33. Kadioglu S, Kadioglu YK and Akyol AA. Monitoring Buried Remains with Transparent 3D Half Bird's-eye View of Ground Penetrating Radar Data in the Zeynel Bey Tomb In the Ancient City of Hasankeyf - Turkey, J. Geophys. Eng. (Special Issue on Cultural Heritage) 2011; 8 S61-S75.

34. Aslanapa O. Turkish Art and Architecture. New York: Praeger, 1971.

35. Basgelen N. A Unique Global Heritage Sstarting to Count its Remaining Days: Hasankeyf and Tigris Valley. Archaeology / Monument-Environment (Arkeoloji /Anit-Çevre) 2006; September-October 17 p114-119.

36. Eskici B, Akyol AA and Kadioglu YK. Material Analyses and Conservation Problems of the Hasankeyf Zeynel Bey Tomb. Journal of Turkish Archaeology and Ethnography 2008; 8, 15-37. (in Turkish)

37 Uluçam. A. Excavations of Hasankeyf from the Past to Nowadays, Spatial Issue of Konya Book, December 2007. p.681-710. (in Turkish)

Chapter 10

Photogrammetric Error Sources and Impacts on Modeling and Surveying in Construction Engineering Applications

Fei Dai*, Youyi Feng and Ryan Hough

Department of Civil and Environmental Engineering, West Virginia University, P.O. Box 6103, Morgantown, West Virginia 26506-6103, USA

ABSTRACT

This paper reviews photogrammetric error sources and their impacts on modeling and surveying for construction quantity takeoff, quality control, and site safety monitoring applications. These error sources include camera internal parameters (i.e., type, principal point, principal distance, and camera lens distortion coefficients), imaging settings (i.e., shooting distances, baselines, percentage of photo overlaps, number of overlapping photos, camera intersection angles, and angles of incidence), and processing software programs. To augment the body of knowledge on photogrammetric modeling errors, this paper further conducts experiment, which concerns characterization of the behavior of different strategies in selecting reference lines for fixing absolute scale of photogrammetric models. In construction photogrammetric surveying, it is imperative to convert the relative scale of a 3D model into absolute measurements so geometric measurements can be taken. Previous work suggests this can be done through the determination of a reference line in absolute units; however, the position and quantity of reference lines has not been investigated. This experiment attempts to tackle this issue. The result shows that one horizontal reference line in the middle of the object performed with consistent accuracy, but if a specific area on the object needs more accurate measurements, it is best to select a reference line in that area. The review and the experimental findings may help construction professionals better understand the performance of the photogrammetric surveying and apply it in their real-world projects.

INTRODUCTION

Photogrammetry, also called image-based reconstruction or image-based modeling, is an old but powerful technology in many engineering applications. In construction, it has proven to be safe, inexpensive, and efficient to collect as-built geometric spatial data (i.e., position, size, shape, and scale) of building objects and civil infrastructure (Luhmann et al. 2006; Brilakis et al. 2011; Dai and Lu 2013).

Photogrammetric modeling is the process of extracting information from images captured by digital cameras (Blachut and Burkhardt 1989). In the field image collection, camera stations are set up to take the images. However, there are several unknown factors that can affect the accuracy of the results, including camera principal point and principal distance, camera lens distortion coefficients, and imaging settings (i.e., shooting distance, baseline, percentage of photo overlaps, number of overlapping photos, camera intersection angles, and angles of incidence). In comparison to controlled settings such as manufacture shops and laboratories, construction sites pose many practical constraints such as the desirable distances and angles to place the camera not being accessible or some details of the target object being obstructed by other objects (e.g., supporting rigs). These practical constrains that exist may lead to the general guidelines of applying photogrammetry not as efficient as they are used in controlled settings. It is also desirable to have site engineers master a bit the fundamentals as well as factors that affect performance of the results so that they can best utilize this technology in construction engineering applications. To this end, this paper makes a comprehensive review of research that has taken place on evaluation of these factors.

The modeling process by photogrammetry can be materialized through three main different ways, including combined special forward-backward intersection, relative-absolute orientation and bundle adjustment methods. All these methods apply the collinear equations which can be set up based on the corresponding points in the images and the real world to calculate the 3D model coordinates of the object. Nevertheless, since the absence of the real world coordinates or an already known field distance in the modeling process, the recovered modeling space geometry as well as the object model itself has an arbitrary relative scale. Thus, it is imperative to convert the relative scale of a 3D model into absolute scale so geometric measurements can be taken. To augment existing knowledge on photogrammetric modeling errors, experiments are conducted to investigate the effects of several different factors on the accuracy of the model absolute scale fixing process.

The review and experimental findings provide a holistic view of how photogrammetric surveying is impacted by different factors. This is particularly helpful as the construction site always features chaos and practical constraints such that the review and experimental findings from this research may help engineers better understand the performance of the photogrammetric surveying and therefore better plan their measurements in field.

REVIEW

The review presented here outlines current practice of construction quantity takeoff, quality control, and site safety monitoring, and recent research efforts on evaluating photogrammetry with its settings for construction surveying related applications. These two are followed by an overview of the fundamental knowledge in the systematic errors due to the camera factors and poor planning of camera network geometry.

Current Practice of Construction Quantity Takeoff, Quality Control, and Site Safety Monitoring

In the construction industry, accurately taking geometric measurements of onsite elements and products is a critical task in almost everyday work to ensure site structural integrity, personnel safety, and timely project delivery. Geometric measurements can be in the form of spatial position, dimension, quantity, and displacement that are typically used in site activities such as quantity takeoff, quality control, and structure safety monitoring.

Quantity takeoff is the process of taking measurements about counts, lengths, areas, volumes, and conditions of a site project for the purpose of quantifying workloads required for completing a certain task. A typical example is quantifying painting areas of interior and exterior walls of a built structure in order to prepare a proposal for bidding. Traditionally, these can be done with a measurement tape, a total station, or a laser scanner depending on the scale and accessibility of the measuring area. However, painting or repainting a building is usually considered a small project. Cost for performing such measurement with an expensive tool such as the total station or laser scanner can hardly be justified. Also, considering that the large scale or inaccessible areas of the building sometimes exist, it is not always practical for the contractors to perform such a job using a traditional measurement tape.

Another class of site surveying activities can be categorized as quality control. It concerns obtaining geometric measurements of positions, dimensions, and sizes of an as-built product first, and then comparing the obtained values with as-designed/planned ones so as to ensure that the specifications are met. For instance, on construction site, engineers have the needs to check dimension quality of prefabricated housing units (i.e., precast façade, Figure 1), which are widely used in constructing tall buildings in dense urban areas. Assuring the dimension quality is vital to successful and timely installation of façades on the construction site. However, current practice largely relies on manual measurement of the geometric dimensions of each façade unit by tapes. To be cost effective, such operations only apply to a limited number of specimens randomly selected from a large amount of façade products.

Figure 1: Precast façade installation on site.

The site safety monitoring involves continuously or periodically evaluation of deformations of a permanent (e.g., building) or temporary (e.g., scaffolding) structure that usually bears dynamic excitements. The consequences of site safety monitoring failure are usually severe, and can cause damage, injuries, or even casualties. A specific case is monitoring the settling displacements of an existing building adjacent to a jobsite. For construction sites, it is stringently demanded that the settlement of adjacent buildings is monitored in the course of foundation excavation to ensure the integrity of these buildings and minimize disturbances. Traditionally, the settling markers are installed on the wall of a monitored building, and the contractor employs a registered professional surveying team to carry out periodical (e.g., bi-weekly) measurements of the settling markers with a total station during the entire piling construction period (ASD 2007). In such respect, surveying settlement consists of two phases: transiting the geodetic coordinates from the government benchmark positions to locally defined points around the monitored building, and performing a periodical surveying of the vertical movements of these markers on the building relative to the local points and identifying changes, both of which remand a sheer volume of efforts and costs.

From the above, it can be easily noticed that applying tape measurement is simple but slow, tedious, and unsafe. While laser scanners (e.g., Jaselskis et al. 2005; Akinci et al. 2006; Bosché et al. 2009; Tang et al. 2011) or total stations (e.g., Shen et al. 2011; Al-Ali 2011) are accurate, fast, and very little training is required for the surveyors, these technologies suffer from high equipment cost. For small projects, they are impracticable since the projected savings hardly justify employing such devices. Also, the equipment is not easily portable and time spent for on-site data surveying is another issue that impedes the application of these technologies. In contrast, photogrammetry is a technology that utilizes cameras or camcorders to collect spatial data of the environment in the form of point clouds or wireframe representations. It has the advantages of low equipment cost, fast on-site data acquirement, and high portability, therefore, holding great potential to allow engineers to rapidly, accurately perform the site surveying in a safe and cost-efficient way.

Evaluating Photogrammetric Modeling with its Settings

While promising, to cater to the needs for taking geometric measurements on site objects to solve site surveying related problems, it is vital to address the accuracy issue of photogrammetry. In the other words, it should be ensured that the achievable accuracy level of this technology matches up to the desired accuracy level for a particular application.

To this end, extensive research efforts have been undertaken that primarily focus on the evaluation of photogrammetry in collecting spatial data of infrastructure elements. Baltsavias (1999) evaluated digital photogrammetry with respect to Digital Terrain Model (DTM) and Digital Surface Model (DSM) generation. At the time the study was conducted, this technology was relatively new. Thus, the focus of the evaluation was placed on ease of installation of the platforms in airplanes, automation of measurement, and accuracy and density of the DTM/DSM. Close range performance of this technology was not explored. González-Jorge et al. (2011) evaluated the performance of photogrammetric systems by employing three different cameras (Canon 5D with 14-mm lens, Nikon D200 with 20-mm lens, and Jai BB500GE with 8-mm lens). The authors concluded that the manufacturing quality of camera models can influence the measurement accuracy [e.g., the higher signal to noise ratio (>50 dB) for the Jai camera compared to that (23 dB) of Canon was evidently related to the higher performance of the Jai camera]. Also, the importance of camera calibration was emphasized to enable photogrammetry for measurement related tasks. In addition to these, Remondino and Fraser (2006), and Luhmann (2009) measured impacts of focal length changes. In their studies, the trends of the measurement accuracy with respect to focal length, image scale, and shooting distance were analyzed. Rieke-Zapp et al. (2009) assessed mechanical stabilization of non-photogrammetric cameras (e.g., Canon 5D, Nikon D3, D2X, D80, D200) and the test result showed that the accomplished accuracies, though limited in comparison with professional photogrammetric cameras, had great potentials for a large quantity of close range applications.

In the area of civil engineering, Zhu and Brilakis (2009) evaluated photo/videogrammetry, active optic triangulation, and laser scanning techniques based on civil infrastructure application requirements. Spatial accuracy, automation of acquisition, and portability of devices were evaluated in their research. Nonetheless, the evaluation was only conducted on a qualitative level. Golparvar-Fard et al. (2011) evaluated accuracy of photogrammetry and laser scanning using a masonry block and a site column as samples. Quantitative comparison was conducted in terms of cost, level of automation, accuracy, and ease of use. However, the accuracy metric was tested in very close ranges (less than 10 m) and factors such as distance, resolution, and type of camera model, all of which could affect the resulting accuracy, were not fully investigated. To ameliorate this, a recent study led by Dai et al. (2013) examined the performance of photogrammetry for reconstructing civil infrastructure objects at collection ranges of 5 ~ 50 m. In this study, large-scale infrastructure objects (bridge, building) were selected, and prevailing software packages were tested. A plethora of factors affecting the performance of accuracy, data density, cost, and time were analyzed, including distances, camera models, resolutions, features (texture), and focal lengths, and the accuracy level achievable was found to be at an average level of 6 ~ 10 cm. Similar conclusions were reported in (Klein et al. 2012) and (Bhatla et al. 2012), based on which Klein et al. (2012) verified the feasibility of image-based as-built documentation for operations and maintenance of buildings, whereas Bhatla et al. (2012) concluded that although photogrammetry has great potential, for higher accuracy applications like defect detection, this technology in its present state is not the most appropriate.

Table 1 summarizes the general behavior of photogrammetry with regard to its performance influencing factors based on the literature above. From the table, the measurement errors associated with photogrammetry can be attributed to two large categories: 1) the systematic error due to camera factors, and 2) the systematic error due to poor planning of camera network geometry, both of which will be discussed in detail in the following sections. Note that random errors to human factors are excluded from the discussion below for which the readers can refer to (Dai and Lu 2010) for detail.

Table 1: Behaviors of photogrammetric modeling

Behaviors of photogrammetric modeling	
Factor	**Behavior**
Distance	Farther distance, lower accuracy (40 m: 6 ~ 10 cm)
Number of overlapping photos	The larger number of overlapping photos, the better accuracy
Intersection angle	The closer to 90°, the better accuracy
Angle of incidence	The closer to 0°, the better quality of images, so the better accuracy
Camera model	Higher end, higher accuracy
Resolution	Strong correlation with accuracy (raise by 1.6% per 1 MB increase)
Features (texture)	More features, higher accuracy; higher accuracy by colorful texture than grey/plain texture
Focal length	Longer focal length, higher accuracy; set to obtain optimum coverage
Camera lens	Less distortion, higher accuracy

Systematic Error Due to Camera Factors

In measurement, the systematic error is a bias that occurs consistently (Viswanathan 2005), resulting from inaccuracy of a system as one observes or measures. For the image-based system, the systematic error due to camera imperfection can be categorized into three aspects: 1) lens distortion, 2) approximated principal distance, and 3) resolution.

Measurement errors due to camera lens distortion are illustrated in Figure 2. The lens distortion makes an image point on the image plane shift from its true *p* (*xn, yn*) to a disturbed position *p'*(*xn', yn'*), resulting in an offset between the two positions. Denoting the offset by *dx* and *dy*, the true coordinates of any image point can be compensated by $xn = xn' + dx$ and $yn = yn' + dy$. For modern digital cameras, the camera lens distortion (i.e., *dx* and *dy*) can be taken as the aggregate of the radial distortion and the decentering distortion (Beyer et al. 1995; Fraser 1996a). As the lens of a camera is actually composed of a combination of lenses, the centers of those lens elements are not strictly collinear, giving rise to decentering distortion. In contrast, the radial distortion occurs in each single optical lens and the distortion effect is magnified along the radial direction of the lens: the further a point is away from the center of the lens, the larger error is produced for its projected image point. Therefore, *dx*, *dy* can be decomposed by $dx = dxr + dxd$and $dy = dyr + dyd$. Assuming that the optical axis of the lens is perpendicular to the image plane, a rigorous and accurate modeling of the distortion can be formulated (Brown 1971) as:

$$
\begin{aligned}
dx_r &= K_1(x'_n - x_p)r^2 + K_2(x'_n - x_p)r^4,\\
dy_r &= K_1(y'_n - y_p)r^2 + K_2(y'_n - y_p)r^4\\
dx_d &= P_1\left[r^2 + 2(x'_n - x_p)^2\right] + 2P_2(x'_n - x_p)(y'_n - y_p),\\
dy_d &= P_2\left[r^2 + 2(y'_n - y_p)^2\right] + 2P_1(x'_n - x_p)(y'_n - y_p)\\
r^2 &= (x'_n - x_p)^2 + (y'_n - y_p)^2
\end{aligned}
\tag{1}
$$

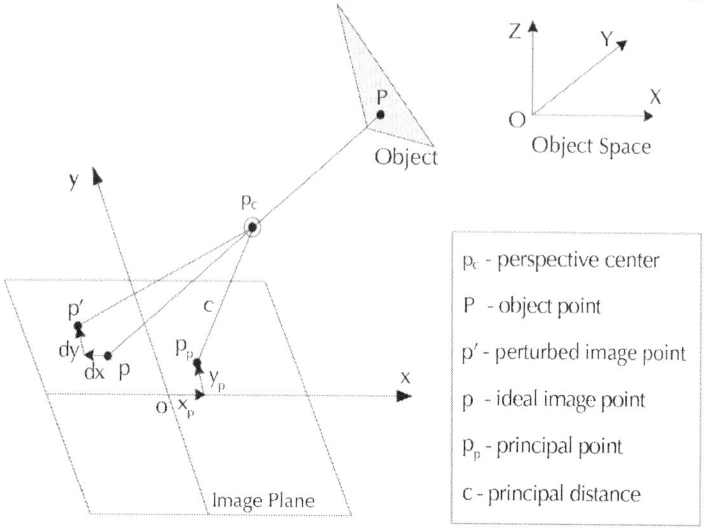

Figure 2: Interior orientation parameters.

Here *xp* and *yp* are the coordinates of the principal point, *K1*, *K2* and *K3* are the radial distortion parameters, and *P1* and *P2* are the decentering distortion parameters. The principal point (*xp*, *yp*) can be referred to as the projected position of the light ray on the image plane through center of the lens opening of the camera (perspective center) from infinity (Figure 2). The model (Eq. 1) can be extended by two further parameters to account for affinity and shear within the image plane, but such parameters are rarely if ever significant in modern digital cameras. Despite this, because focuses have been made upon 3D modeling and visualization, most of the current 3D reconstruction systems, when developed, actually do not take the lens distortion into account and simply assume $x_n' = x_n$ and $y_n' = y_n$ (e.g., Snavely 2010; Pollefeys et al. 2008). However, this is hardly acceptable when they are applied to surveying related tasks, especially when working with cameras that are not designed for professional photogrammetric needs. Such scenario is common in close range applications (Rieke-Zapp et al. 2009). The mechanical stability of the off-the-shelf digital cameras is usually poor in comparison to high-accuracy photogrammetric requirements (Luhmann 2010). As a result, it is indispensable for the reconstruction process chain to incorporate correction of camera lens distortion (Shortis et al. 1998).

The principal distance (*c*) of a camera in photogrammetric modeling is defined as the distance of the perpendicular line from the perspective center (center of lens opening) to the image plane of the camera (Figure 2). It is often approximated to be the camera's focal length (*f*) in a reconstruction process on the grounds that the focal length of a camera is easy to access, for example, through camera specifications or by referring to information stored in the exchangeable image file (EXIF) format of the captured images. However, this can also lead to the systematic errors of the measurement. Figure 3 shows a camera photographing an object where the photographic object distance and the image distance are illustrated. The principal distance c equals the image distance v when the image plane is at the exact position along the optical axis and an object is clearly focused. Meanwhile, the distance between the object and the camera lens is defined as object distance u. The conjugated distances u, v and the focal length f are related by the lens conjugate equation (Ray 1984) as $1/u + 1/v = 1/f$, by which the image distance can be derived from $v = uf/(u - f)$, leading to the actual length of the principal distance. In reality, as an object is shot, the object distance u is usually much farther than the image distance v. As such, the denominator $(u - f)$ can be approximated as u, which yields: $v \approx f$. This explains why the principal distance (*c*) can be approximated as the focal length of the camera lens (*f*). However, the principal distance (*c*) exactly equals the focal length (*f*) when and only when focused at infinity, namely, $c \approx f$; otherwise the principal distance should be corrected to improve the measurement accuracy.

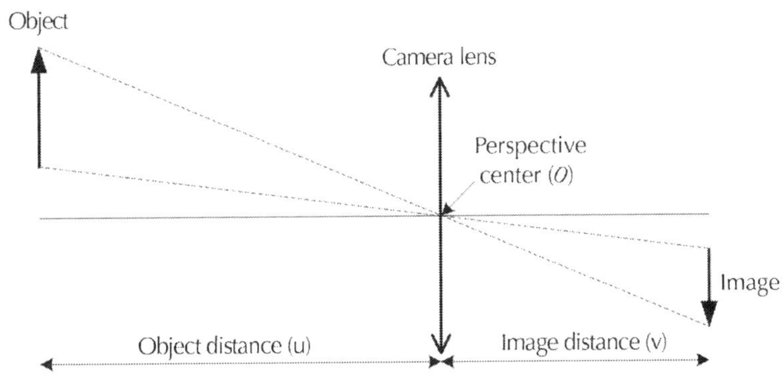

Figure 3: Photographing illustration.

The systematic error due to the above two factors can be minimized by adjusting their parameters through camera calibration. Camera calibration is the analytical procedure of determining the camera's internal parameters including the principal distance, format size, principal point, and lens distortion coefficients. Commonly adopted methods for performing this work include Zhang (1999), Heikkilä and Silvén (1997), Tsai (1987), Rüther (1989), Fraser (1997), Gruen and Beyer (2001), Bouguet (2012), PhotoModeler (Eos System Inc 2012), and CameraCalibrator (Photometrix 2012). These methods adopt a combination of linear and non-linear solving strategies where a linear method (e.g., Direct Linear Transformation) is used to recover initial approximations for the parameters, followed by a non-linear algorithm (e.g., Levenberg-Marquardt) in search of refined solutions of the orientation and calibration. Using a fully calibrated camera in the measurement of a 3D object can yield the best results.

In addition, the maximally allowed resolution can also limit the reconstruction accuracy. Given a fixed focal length and picture taking distance, the resolution (i.e., number of pixels) of a camera determines the smallest size of items that can be precisely located on the camera's image sensor (CMOS or CCD), namely, how small an item can be carried by a pixel on the sensed image. This provides a theoretical cap of accuracy for this image-based technology, and explains why a high-resolution camera is favorably recommended for a photogrammetric system (Dai et al. 2011). The higher the resolution of a camera, the better chance is there to make the highest accuracy achievable.

Systematic Error Due to Poor Planning of Camera Network Geometry

The systematic error that takes place in a photogrammetric measurement can also result from poor planning of camera network geometry. In photogrammetry, camera network geometry is denoted as the imaging settings specifying the shooting distance, baselines, percentage of photo overlaps, number of overlapping photos, camera intersection angles, and angles of incidence such that an accurate 3D scene can be reconstructed. In the fields of surveying and civil engineering, researchers investigated metrological behavior of photogrammetry by evaluating the impact of individual settings of the network geometry on accuracy when being applied for civil infrastructure scenes.

Initially, most testing for photogrammetry focused on evaluating the behavior of the multi-image setup for the camera calibration (Luhmann et al. 2006). Images were manually recorded either for a test field or for the measured object itself, both of which required a favorable network geometry in which convergent and rotated images of a preferably 3D object should be acquired, with well distributed points throughout the image format (Remondino and Fraser 2006; Luhmann et al.2006). According to (Fraser 1982), in a geometrically weak network, correlations may lead to instabilities in the least-squares adjustment for estimating the camera interior parameters, which in turn could undermine the use of the estimated results in the bundle adjustment solution. Several experimental studies in close-range photogrammetry have revealed that the accuracy of a network increases with the increase of convergence angles of cameras, the numbers of intersection rays to an object point, the number of measured points, and orthogonal roll angles (Fraser 1996b; Fryer 1996; Clarke et al. 1998; Gruen and Beyer 2001; El-Hakim et al. 2003; Remondino and Fraser 2006). Here the orthogonal roll angles require that at each location where an object or test field is shot, at least two images are with ±90° (portrait and landscape). This is particularly important when applying photogrammetry to surveying related tasks, as failure of doing so can break coupling between parameters of the interior and exterior orientations. A counterexample was demonstrated by Remondino and Fraser (2006), in which a camera was moved through various orientations but without any notable 90° rotations. As a result, the calibration procedure led to a very high coupling result between the interior and exterior orientation parameters. In this case, the variations of the principal point and distortion parameters did not affect the root-mean-square error (RMSE) of the camera calibration residuals. It was highly probable to produce low accurate calibration results and high systematic errors of the measurement.

Recently, the use of photogrammetry for reconstructing large-scale infrastructure scenes has motivated a flurry of research in testing the accuracy of camera network geometry for field measurement (Wackrow and Chandler 2008; 2011; Dai and Lu 2008; 2010; 2013; Dai et al. 2013). These studies manifested that the geometry of camera network, if not meticulously planned, can also lead to the systematic error of the measurement. A number of settings have been studied to reveal the trend and level of the errors by which the accuracy can be estimated. In a series of tests, Wackrow and Chandler (2008) identified significant dependency of the systematic error on the lens model, and imaging configurations. According to (Wackrow and Chandler 2008; 2011), in contrast to the normal case in which the camera base is parallel to the object plane and the optical axes of the cameras intersect the object plane orthogonally, using mildly vertical or oblique convergent imagery for terrain elevation modeling can minimize the remaining systematic error caused by slightly inaccurately estimated lens distortion parameters. In terms of camera intersection angles, the work of modeling construction site objects using photogrammetry (Dai and Lu 2008) revealed that the closer the angle between the light rays is to a right angle (90 degree), the smaller error would occur (Figure 4). The work was then extended by (Dai and Lu 2010) to assess the applicability of photogrammetry in taking geometric measurements on building products. A very promising result was achieved as the measurement accuracy in the test was at the level of 1 cm. From this work, it was also found that by empirically choosing the number of overlapping photos, intersection angles, angles of incidence, and a deliberate calibration, a higher level of accuracy could be achievable. Nevertheless, the impacts of the imaging settings were not addressed in the study. To fill this gap, a 3D modeling method based on the mechanism of photogrammetry was created, which served as a test-bed to evaluate the impact of the baseline, the overlap percentage, and the quantity of overlapping photos on the resulting accuracy of photogrammetry (Dai and Lu 2013). In this work, camera stations were set parallel to each other in order to achieve maximum coverage of the scene. Among the results, one was particularly noteworthy: there is always a cut-off point between the length of baseline and percentage of overlap that makes the modeling achieve the best accuracy while utilizing the minimum number of photos. At the shooting distance of 15 m, 4 m baseline achieves the best accuracy of the planarity with the photo overlap percentage of 55% (Figure 5). Regard to other shooting distances (25, 40, and 55 m), the recent work of (Dai et al. 2013) measured the accuracy and the results shed light on a strong correlation of the measurement error with the shooting distance. Based on the results (Dai et al. 2013), increasing the shooting distance even by 1 m can lead to a decrease of the measurement accuracy as much as 2.5%.

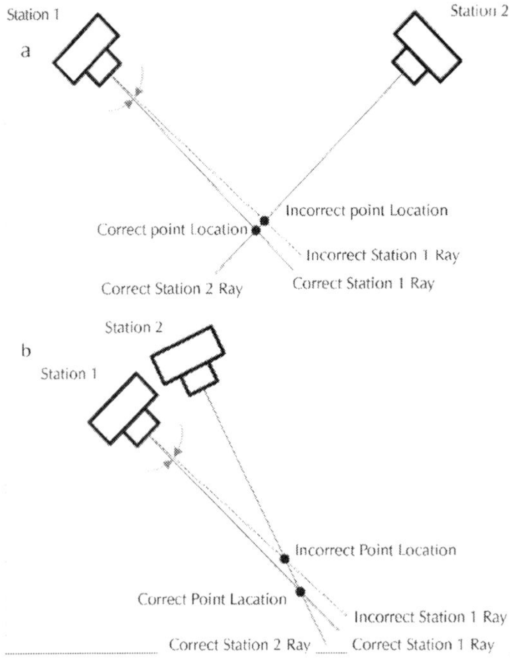

Figure 4: Point location error with a) good and b) bad camera positions (Dai and Lu2008).

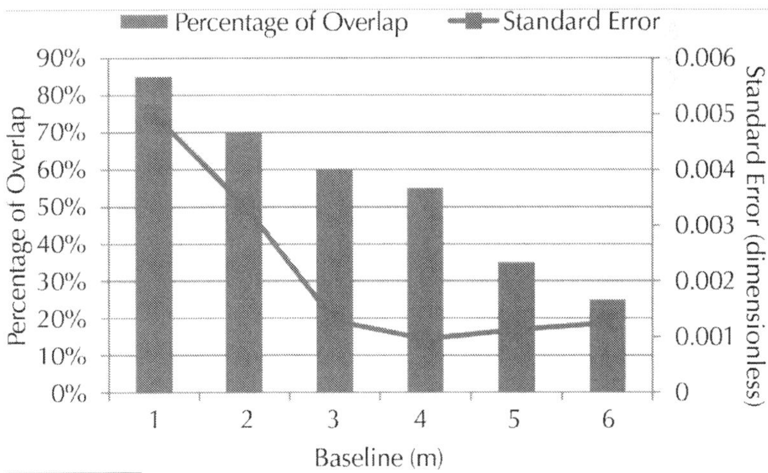

Figure 5: Impact of baseline and percentage of overlap (Dai and Lu2013).

FIXING ABSOLUTE SCALE OF PHOTOGRAMMETRIC MODELS

In addition to the existing accuracy influencing factors discussed above, another common issue involved the photogrammetry modeling process for construction engineering applications is the model scale fixing process. To extract some useful data from a photogrammetric model such as structure displacement and deformation, engineers need to perform the geometric measurements of the model. Before the scale fixing process, the model only has a relative scale which is is not sufficient. As a result, it attaches great significance to explore the model scale fixing strategies and the accuracies as an addition to the above review.

In fact, there are several different methods available to do this in various photogrammetry applications. Control points are widely used for setting the absolute orientation (e.g., Saidi et al.2011). They are points with known coordinates that can be referenced to set the absolute scale of a model. In aerial photogrammetry surveying, ground control points (GCPs), which are identifiable object with a known size, are selected so that the scale of the model can be set in real world units. This method is limited when the terrain is mountainous and covered with foliage and trees and if there is a lack of clear existing usable targets. Also, GCPs are difficult to set in a convenient and computer compatible form (Smith and Park 2000). In general, GCPs are not readily available in most situations, making this method impractical. A method using GPS/IMU can simultaneously measure the full exterior orientation while data is being recorded. This method involves two steps: (1) GPS/IMU pre-processing and (2) pre-determined sensor calibration. In the first step the GPS signal and IMU measurements are transformed into object space coordinates. There are two methods that Wegmann (2002) suggests for sensor calibration, a direct or integrated method, both of which requires GPS/IMU observations and ground control information. This method produces precise results for various applications; however, it is not widely available and can be tedious. Using a stereo rig that fixes the baseline of two cameras is another way to enable absolute measurement of an object. This method requires pre-configured camera set and maintaining two cameras' relative position when taking photos in field (Brilakis et al. 2011). As the baseline is known, the relative and absolute orientations can be accomplished simultaneously. The stereo photogrammetry is often used, but extra efforts and time is needed for preparing the rig and camera set. In addition, a premeasured cube with known dimensions is used to calibrate the absolute scale of the generated photogrammetric models (Rashidi et al. 2013).

Out of all the techniques for absolute scale fixing, applying reference lines is chosen in our study considering that it is most time-effective, easy-to-use, and performs with sufficient accuracy for construction applications.

EXPERIMENT DESIGN

The following section describes the design and technical components of the experiment. This process is quantified such that it could easily be implemented by any construction professional that has the necessary equipment.

Calibrating Camera/Lens Parameters and Software

To produce the models for our experiment, two cameras were used: a Canon EOS Rebel T3i and a Canon EOS 60D, and two lenses were used: a Canon EF 35 mm and a Canon EF-S 18-135 mm. These camera lenses are made up of a combination of lenses all assumed to be collinear. The photogrammetry software that was used was iWitness, which is one of the leading close-range photogrammetry software systems because of its accurate 3D measurements and usability. To begin creating models in iWitness, first the cameras used must be calibrated. The calibration process corrects the radial and decentralizing lens distortion so that an accurate coordinate system and accurate scales can be established. In our research, we applied a fully-automated process and a self-calibration process in iWitness (Fraser 1997) to ensure that any error due to the camera lens would be minimized. Note that although this research employed iWitness, other software packages such as PhotoModeler can also be used, which are expected to produce the similar results since the fundamentals of these program packages are the same.

Photo Taking and Setting Scale Bar Scenarios

After the cameras were calibrated, photos were taken of each object from an evenly distributed quantity of positions and angles around the surrounding area of the object. To ensure accuracy, it is a good practice to cover each point on the object in three or more photos. Red-reflective markers were attached to the object so that iWitness could recognize the spatial characteristics and establish the relative orientation of the object. This is especially useful when the object being modeled does not have prominent natural features to manually reference. A combination of red-reflective points and natural features on the object can be used to set an accurate relative orientation. The photos were taken while different quantities and locations of scale bars (Figure 6) were placed in the scene. The scale bars used in the experiment were finely precalibrated to be 30" and were used to convert the relative orientation to absolute units (Figure 7). Six scale bar scenarios (Figure 6) were carried out on each object and the scale bars were kept from one another with relatively large distances. This is because using a predefined small dimension for calculating the absolute scale of a large scene may result into errors of large magnitudes. This experiment would observe the impact of this drift problem. The scale bar quantity and placement would be characterized and strategies would be given to construction professionals. For walls and other vertical objects, scale bars should be placed on tripods so that they can be seen clearly, but for objects on the ground (e.g., sidewalks, parking lots), scale bars can be placed directly on the surface of the object.

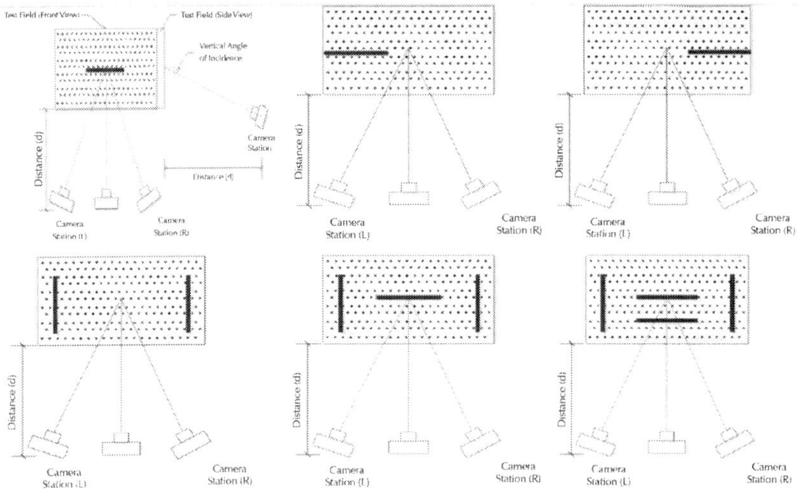

Figure 6: Six scale bar scenarios (top row from left to right: scenario 1, scenario 2, scenario 3; bottom row from left to right: scenario 4, scenario 5, scenario 6).

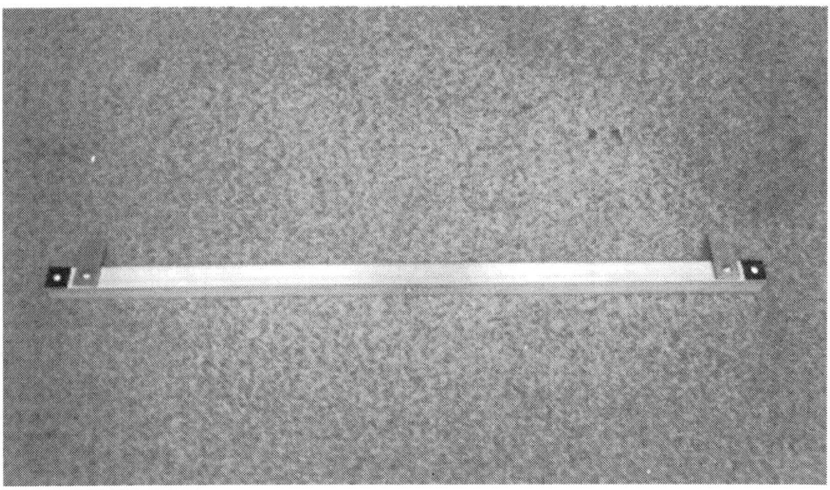

Figure 7: Scale bar.

Marking, Referencing and Setting Scales of Models

Once the photos were taken of each object, they were imported into iWitness. After the camera were adequately calibrated and the project photos were imported, points could be marked and referenced. Point marking and referencing is a process that matches the same points in different photos, taken from different locations and angles, so that the relative orientation of the object can be established. As mentioned earlier, several red reflective markers were taped to the objects so that this process would be accelerated by the automated feature in iWitness. After the relative orientation of the model was established by the marking and referencing process, the absolute orientation was set by setting the scale bar distance (or distances) to its 30″ calibrated length. Once the scale bars were referenced in each model, accuracy comparisons were made on the models.

Accuracy Comparisons

There is a built-in distance function in iWitness that allows the user to measure the distance between two points. After the scale bars were referenced in the individual models, twenty or more sample distances were determined so that there would be a significant sample size to do a comparison. The sample distances for each object were collected by a total station or a tape measure and were compared to the distance values given by the iWitness models produced by the six scale bar scenarios. The values were compared to one another in hopes to characterize scale bar placement.

EXPERIMENTAL IMPLEMENTATION AND RESULTS

In implementing the experimental procedure for the selected objects, photos were taken from approximately two meters away. This ensured that a high level of accuracy would be accomplished. The three objects that were modeled were a section of sidewalk, a section of a retaining wall, and a small storage building (Figure 8). These objects were selected so that there would be a variety of object types (horizontal planar shape, vertical curved shape, and vertical planar shape) and so that biases would be eliminated.

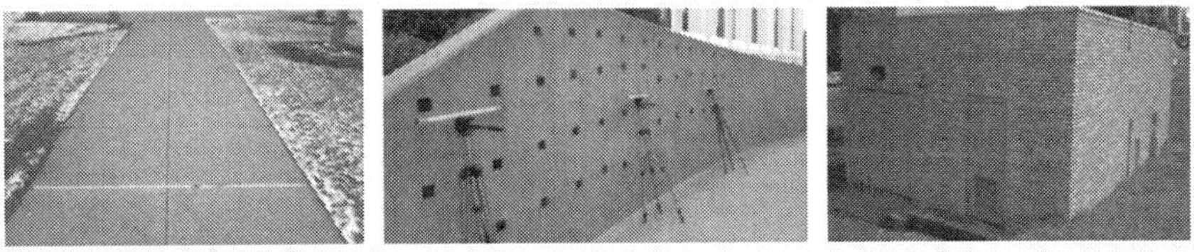

Figure 8: Left to right: sidewalk slab, retaining wall, and storage building.

Table 2 summarizes the settings of each model, including the combination of camera and lens, the number of photos, the baseline of two consecutive camera stations, the percent coverage which is the overlapped area between two adjacent photos, and the resolution of the photos.

Table 2: Settings of each model

	Slab model	Retaining wall	Full building
Camera Model	Canon EOS Rebel T3i	Canon EOS 60D	Canon EOS 60D
Lens Model	Canon EF-S 18-135 mm	Canon EF 35 mm	Canon EF 35 mm
# of Photos Taken	82	85	76
Baseline (m)	1	1	1.5
Percent Coverage	50%	58%	43%
Resolution (mp)	18	18	18

Object 1: Sidewalk Slab

The concrete slab model presented a challenge because of its position on the ground and its planar shape. The scale bars were laid directly on the ground In the scenario positions listed above. There were twenty sample distances taken between two corners of individual rectangular sections within the slab, and these sample distances were oriented diagonally, vertically, and horizontally to ensure that accurate comparisons could be accomplished. In order to take sample distances in this manner, these points had to be manually referenced in iWitness on three or more photos for sufficient accuracy. The ground truth values, compared against the distances given by the six models, were collected by a tape measure.

Overall, results from all scenarios are comparable. Scenario 2 performed with the best accuracy results with an average absolute difference of 3.25 mm, an average percent difference value of 0.18%, and a standard deviation of 2.76 mm. Scenario 5 was next best (3.36 mm, 0.18%, and 3.41 mm), then Scenario 6 (3.51 mm, 0.19%, and 3.40 mm), Scenario 4 (3.54 mm, 0.19%, and 3.02 mm), Scenario 3 (3.58 mm, 0.19%, and 2.76 mm), and lastly Scenario 1 (4.18 mm, 0.21%, and 3.22 mm). Figure 9 shows a scatter plot of the relative error values by Scenario. The majority percent differences are within 0.4%.

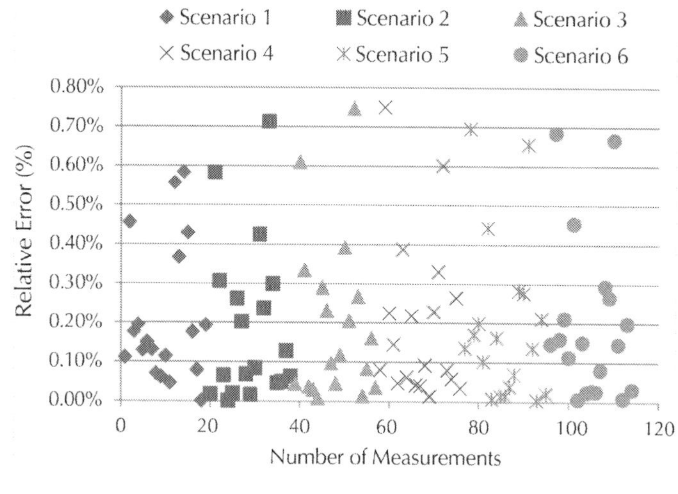

Figure 9: Relative errors by scenario for slab measurements.

Object 2: Retaining Wall

The curved shape of the retaining wall presents a difficult task when attempting to produce accurate models. The scale bars were attached to tripods and arranged according to the scenario. Thirty-two sample distances were selected between the red reflective points on the markers attached to the wall and then used for accuracy comparisons. The sample distances were selected in this way because there were no prominent natural features on the retaining wall that could be manually referenced and used to create sample distances. The ground truth data was collected by measuring the distance between the centers of two points on the markers with a tape measure.

For the left side, Scenario 2 on average performed with the greatest amount of accuracy with an average absolute difference of 2.29 mm, an average percent difference of 0.16%, and a standard deviation of 1.90 mm. Scenario 1 performed with the next best accuracy (3.67 mm, 0.33%, and 1.85 mm); then Scenario 6 (6.79 mm, 0.58%, and 2.56 mm), Scenario 5 (6.83 mm, 0.59%, and 2.56 mm), Scenario 4 (7.63 mm, 0.65%, and 2.71 mm), and lastly Scenario 3 (9.37 mm, 0.75%, and 3.24 mm). These results are interesting because even though Scenarios 4, 5, and 6 have more scale bars referenced on both sides of the wall, Scenario 1 and 2 still outperformed them. Scenario 2 outperformed Scenario 3 which we expected because Scenario 2 has a scale bar referenced on the left side of the wall. Figure 10a shows the percent differences (relative errors) for the left side sample distances.

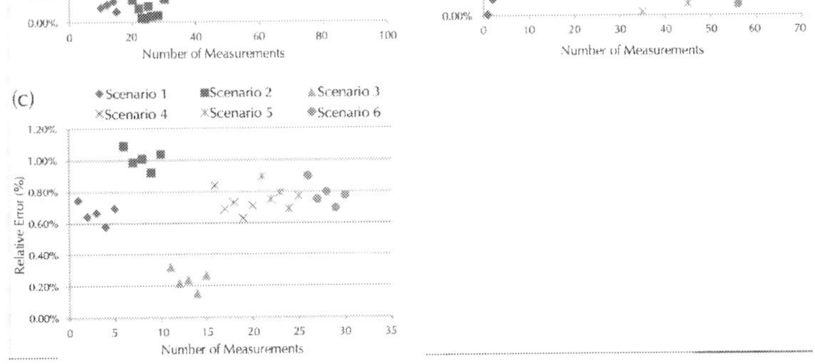

Figure 10: Relative errors by scenario for retaining wall measurements of a) left side, b) middle side, and c) right side.

For the middle section, Scenario 1 (3.60 mm, 0.19%, and 2.26 mm), Scenario 3 (3.49 mm, 0.24%, and 1.34 mm), and Scenario 4 (3.58 mm, 0.19%, and 2.42 mm) performed with the best accuracy. Scenario 5 (4.51 mm, 0.24%, and 2.91 mm) and Scenario 6 (4.56 mm, 0.24%, and 2.94 mm) performed with similar accuracy. Scenario 2 (9.29 mm, 0.53%, and 4.55 mm) performed with the least amount of accuracy. Figure 10b shows the percent differences for the middle side sample distances.

For the right side, Scenario 3 (1.49 mm, 0.24%, and 0.61 mm) performed with the highest level of accuracy; Scenario 1 (4.13 mm, 0.66%, and 1.30 mm) was the next best; Scenario 4 (4.47 mm, 0.72%, and 1.44 mm) was the third most accurate, then Scenario 5 (4.84 mm, 0.78%, and 1.54 mm) and Scenario 6 (4.86 mm, 0.78%, and 1.54 mm) which performed similarly, and lastly Scenario 2 (6.26 mm, 1.01%, and 1.90 mm) performed with the least amount of accuracy. Figure 10c displays the percent differences for the right side sample distances.

Overall, Scenario 1 performed with the highest amount of accuracy (3.72 mm, 0.33%, and 1.89 mm) and then Scenario 6 (5.69 mm, 0.49%, and 2.73 mm) and Scenario 5 (5.69 mm, 0.49%, and 2.74 mm). The prevailing patterns that we saw throughout this model were Scenario 2 performed best on the left side, Scenario 3 performed best on the right side, Scenario 5 and 6 had nearly identical results, and on average Scenario 1 performed had the best results.

Object 3: Storage Building

The storage building model was the only full object model completed in our research. The six scale bar scenarios were carried out on each of the four walls, so, for the model generated under Scenario 1, there were four total scale bars (one placed in the middle of each wall). Twenty-four sample distances were taken in between the corners of individual bricks on the wall. This was chosen because they provided clear points that could easily be manually referenced and compared to ground truth data. Six sample distances were collected on each of the four faces of the building by a tape measure and used for comparison.

Overall, Scenario 3 performed with the highest level of accuracy which had an average absolute difference value of 1.63 mm, an average percent difference value of 0.18%, and a standard deviation of 1.37

mm. Scenario 2 (1.67 mm, 0.18%, and 1.43 mm) was the second most accurate, then Scenario 1 (1.89 mm, 0.20%, and 1.78 mm), Scenario 5 (2.07 mm, 0.24%, and 1.71 mm), Scenario 6 (2.09 mm, 0.24%, and 1.60 mm), and lastly Scenario 4 (2.09 mm, 0.24%, and 1.60 mm). This result is surprising because Scenarios 1–3 outperformed Scenarios 4–6 which have more scale bars referenced. There are similar patterns that were already observed in the previous two models: 1) Scenario 2 generally performed better for sample distances on the left side of each wall (where its scale bar was referenced) and Scenario 3 performed better for sample distances on the right side of each wall, 2) Scenario 5 and Scenario 6 performed with similar results. Figure 11is a scatter plot of the percent differences. The scatter plot is grouped by scenario to display how the scenarios performed over the whole model.

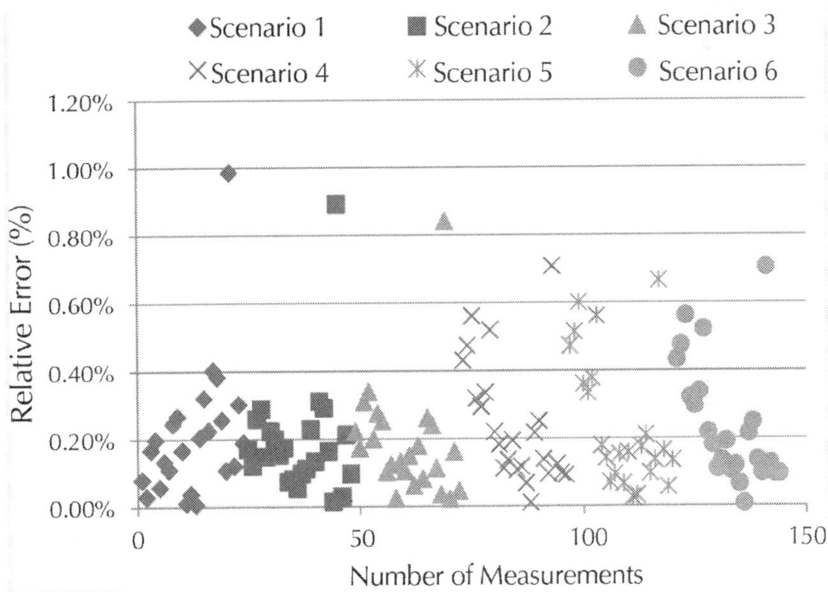

Figure 11: Relative errors by scenario for storage building measurements.

The other data analysis that was carried out on this model was a quantity comparison. Instead of arranging the scale bars in the six scenarios, one scale bar was referenced in the middle of each wall of the building (in other words, four total scale bars) and the sample distance values were obtained on the model. To compare the effect of the quantity of total scale bars, we decreased the number of scale bars by increments of one and recorded the sample distance values.

As expected, the trial with four total scale bars referenced performed with the greatest amount of accuracy having an average absolute difference of 1.72 mm, an average percent difference of 0.19%, and a standard deviation of 1.50. The trial with three total scale bars referenced performed with the next best results (1.89 mm, 0.20%, and 1.78), two total scale bars performed with the next best results (2.24 mm, 0.24%, and 2.05), and lastly one total scale bar performed with the worst results (3.39 mm, 0.35%, and 2.48). Figure 12 shows a scatter plot of the percent difference values that were arranged by quantity.

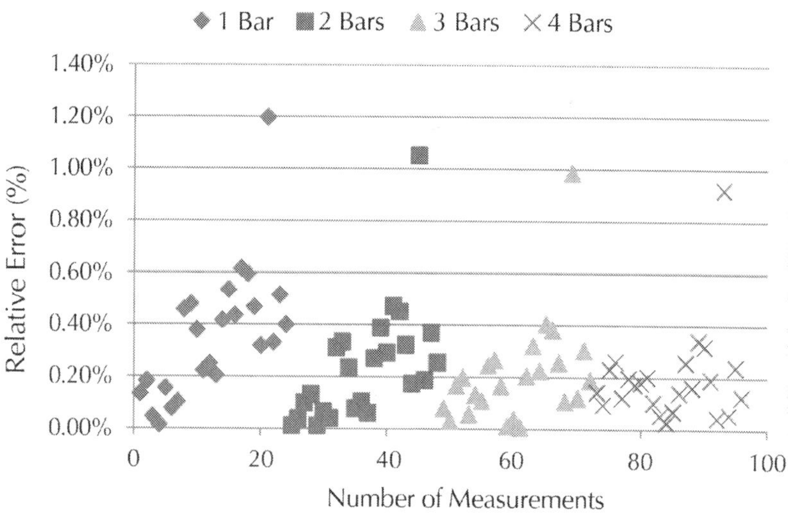

Figure 12: Relative errors by quantity for storage building measurements.

Discussion

There were three prevailing patterns that surfaced throughout: 1) Scenario 1 performed with consistent accuracy results for all sample distance locations (left, middle, right), 2) Scenario 2 was more accurate for distances located on the left side of the model (where its scale bar is referenced) and Scenario 3 was more accurate for distances located on the right side of the model, and 3) Scenarios 5 and 6 performed with similar accuracy results while Scenario 4 performed slightly less accurately in comparison.

For the most consistent results with one reference line, we would suggest referencing it horizontally in the middle of the selected object. If there is a particular area on an object that needs to be modeled more accurately than others, then it would be a good practice to place a reference line in that area also. As far as quantity is concerned, there was a leveling off of accuracy benefit that happened from three scale bars to four scale bars, it suggests that two or three scale bars be referenced on the selected object can achieve the best results.

CONCLUSION

This paper provides a comprehensive review of state of practice in construction quantity takeoff, quality control, and site safety monitoring, and recent research efforts on evaluating photogrammetric error sources and their impact on modeling for construction surveying related applications. In addition, this paper conducted experiments to characterize the behavior of fixing full-scale photogrammetric models in construction applications. In this experiment, factors including relative position of the reference lines, number of reference lines, and spatial relationships of the reference lines were considered. The surveying results were then compared with ground truth data, which allows for accuracy determination and aids in determining the best strategy in selecting reference lines for scale fixing. The review and the experimental findings may help construction professionals better understand the performance of the photogrammetry technology and apply it in their real-world projects.

Based on the existing literature and the experimental results, particularly on the accuracy level achievable, photogrammetry presents potential for modeling of 3D geometry of site elements including building products, equipment, and temporary facilities. These 3D models may facilitate construction management functions such as taking geometric measurements for assessment of quality, quantities and production rates on particular work items, positioning and surveying building elements on site, and cost-effective as-built modeling and visualization of site situations. However, it must admit that photogrammetry is not ubiquitous for modeling of any scenarios. It suffers from difficulties to reconstruct texture-less areas of scenes like walls or window glasses and the inability to generate point clouds of scenes that require high densities. Based on the categories of the error sources - systematic error due to camera factors and systematic error due to poor planning of camera network geometry, it is crucial to take appropriate strategies to minimize the different types of errors. For the systematic error due to camera factors, it is suggested as: calibrating the camera with professional calibration tools before using it in the field; trying to use single-lens reflex (SLR) or better cameras that produce high quality of photos; and using prime lenses instead of zoom lenses whenever possible. For the systematic error due to poor planning of camera network geometry, there are a serious of rules of thumbs that exist such as increasing the overlap of the photo coverage, keeping the baseline as large as possible, and constraining the intersection angles between 60° and 90°. However, the ideal setups based on these rules of thumbs are sometimes hard to be implementable in the field thanks to the practical constrains that exist. Therefore, an automatic guidance system that is field-deployable would be favorable. This will naturally become the extension of the future study. Moreover, an analytical procedure is desirable that is capable of characterizing the quantitative relationships between the data accuracy and the affecting factors (e.g., camera type, image resolution) for achieving a level of desired accuracy. The quantitative relationships will be especially helpful to determine data collection settings for a particular scenario.

REFERENCES

1. Akinci, B, Boukamp, F, Gordon, C, Huber, D, Lyons, C, Park, K (2006). A formalism for utilization of sensor systems and integrated project models for active construction quality control. *Automation in Construction, 15*(2), 124–138.
2. Al-Ali, SH (2011). The practical applications of the electronic total station and computerized system in engineering surveying. *European Journal of Scientific Research, 55*(1), 125–153.
3. Architectural Services Department (ASD) (2007). Particular Specification for Minipiles. Hong Kong SAR.

4. Baltsavias, E (1999). A comparison between photogrammetry and laser scanning. *ISPRS Journal of Photogrammetry and Remote Sensing, 54*(2–3), 83–94.

5. Beyer, HA, Uffenkamp, V, van der Vlugt, G In Gruen A, Kahmen H (Eds.) (1995). Quality control in industry with digital photogrammetry. *Optical 3-D measurement techniques III* (pp. 29–38). Heidelberg: Wichmann Verlag.

6. Bhatla, A, Choe, SY, Fierro, O, Leite, F (2012). Evaluation of accuracy of as-built 3D modeling from photos taken by hand held digital cameras. *Automation in Construction, 28*, 116–127.

7. Blachut, TJ, & Burkhardt, R (1989). Chapter 1. Historical development of photogrammetric methods and instruments. Falls Church, VA: American Society for Photogrammetry and Remote Sensing.

8. Bosché, F, Haas, CT, Akinci, B (2009). Automated recognition of 3D CAD objects in site laser scans for project 3D status visualization and performance control. *Journal of Computing in Civil Engineering, 23*(6), 311–318.

9. Bouguet, JY (2012). Camera calibration toolbox for Matlab. http://www.vision.caltech.edu/bouguetj/calib_doc/ {accessed May 18, 2012}

10. Brilakis, I, Fathi, H, Rashidi, A (2011). Progressive 3D reconstruction of infrastructure with videogrammetry. *Automation in Construction, 20*(7), 884–895.

11. Brown, DC (1971). Close-range camera calibration. *PE&RS, 37*(8), 855–866.

12. Clarke, TA, Fryer, JG, Wang, X (1998). The principal point and CCD cameras. *The Photogrammetric Record, 16*(92), 293–312.

13. Dai, F, & Lu, M In Mason SJ, Hill R, Moench L, Rose O (Eds.) (2008). Photo-based 3D modeling of construction resources for visualization of operations simulation: case of modeling a precast facade. *Proceedings of the 2008 winter simulation conference* (pp. 2439–2446). Miami, Florida: IEEE.

14. Dai, F, & Lu, M (2010). Assessing the accuracy of applying photogrammetry to take geometric measurements on building products. *Journal of Construction Engineering and Management, 136*(2), 242–250. Reston, VA: ASCE

15. Dai, F, & Lu, M (2013). Three-dimensional modeling of site elements by analytically processing image data contained in site photos. *Journal of Construction Engineering and Management, 139*(7), 881–894. Reston, VA: ASCE

16. Dai, F, Dong, S, Kamat, VR, Lu, M (2011). Photogrammetry assisted measurement of interstory drift for rapid post-disaster building damage reconnaissance. *Journal of Nondestructive Evaluation,30*(3), 201–212. Springer, London, UK

17. Dai, F, Rashidi, A, Brilakis, I, Vela, P (2013). Comparison of image-based and time-of-flight-based technologies for three-dimensional reconstruction of infrastructure. *Journal of Construction Engineering and Management, 139*(1), 69–79. Reston, VA: ASCE

18. El-Hakim, S, Beraldin, J, Blais, F (2003). Critical factors and configurations for practical 3D image-based modeling. *The 6th Conference on Optical 3D Measurements Techniques, Zurich, 2,* 159–167.

19. Eos System Inc (2012). PhotoModeler: close-range photogrammetric measurement and 3D modeling software. http://www.photomodeler.com {accessed June 20, 2012}

20. Fraser, CS (1982). On the use of non-metric cameras in analytical non-metric photogrammetry.*International Archives of Photogrammetry and Remote Sensing, 24*(5), 156–166.

21. Fraser, CS In Atkinson KB (Ed.) (1996a). Industrial measurement applications. *Close range photogrammetry and machine vision, 12*, 329–361. Scotland: Whittles.

22. Fraser, CS In Atkinson (Ed.) (1996b). Network design. *Close-range photogrammetry and machine vision* (pp. 256–282). UK: Whittles Publishing.

23. Fraser, CS (1997). Digital camera self-calibration. *ISPRS Journal of Photogrammetry and Remote Sensing, 52*(4), 149–159.

24. Fryer, J In Atkinson (Ed.) (1996). Camera calibration. *Close-range photogrammetry and machine vision* (pp. 156–179). UK: Whittles Publishing.

25. Golparvar-Fard, M, Bohn, J, Teizer, J, Savarese, S, Peña-Mora, F (2011). Evaluation of image-based modeling and laser scanning

accuracy for emerging automated performance monitoring techniques. *Automation in Construction, 20*(8), 1143–1155.

26. González-Jorge, H, Riveiro, B, Armesto, J, Arias, P (2011). Verification artifact for photogrammetric measurement systems. *Optical Engineering, 50*(7), 073603.
27. Gruen, A, & Beyer, HA In Gruen , Huang (Eds.) (2001). System calibration through self-calibration.*Calibration and orientation of cameras in computer vision, 34,* 163–194. Springer Berlin Heidelberg: Springer Series in Information Sciences.
28. Heikkilä, J, & Silvén, O (1997). A four-step camera calibration procedure with implicit image correction. Proceedings of IEEE Computer Society Conference on Computer Vision and Pattern Recognition. 1106–1112.
29. Jaselskis, EJ, Gao, Z, Walters, RC (2005). Improving transportation projects using laser scanning.*Journal of Construction Engineering and Management, 131*(3), 377–384. Reston, VA: ASCE
30. Klein, L, Li, N, Becerik-Gerber, B (2012). Imaged-based verification of as-built documentation of operational buildings. *Automation in Construction, 21,* 161 171.
31. Luhmann, T (2009). Precision potential of photogrammetric 6DOF pose estimation with a single camera. *ISPRS Journal of Photogrammetry and Remote Sensing, 64*(3), 275–284.
32. Luhmann, T (2010). Close range photogrammetry for industrial applications. *ISPRS Journal of Photogrammetry and Remote Sensing, 65*(6), 558–569.
33. Luhmann, T, Robson, S, Kyle, S, Harley, I (2006). Close range photogrammetry: principles, methods and applications. Scotland, UK: Whittles Publishing.
34. Photometrix (2012). CameraCalibrator: software for fully automatic digital camera calibration using coded targets. http://www.photometrix.com.au/ {accessed December 18, 2012}
35. Pollefeys, M, Nister, D, Frahm, J, Akbarzadeh, A, Mordohai, P, Clipp, B, Engels, C, Gallup, D, Kim, S, Merrell, P, Salmi, C, Sinha, S, Talton, B, Wang, L, Yang, Q, Stewenius, R, Welch, G, Towles, H (2008). Detailed real-time urban 3d reconstruction from video. *International Journal of Computer Vision, 78*(23), 143–167.

36. Rashidi, A, Brilakis, I, Vela, P (2013). Monocular videogrammetry for generating absolute-scale 3D point cloud data of civil and infrastructure scenes. Proceeding of ASCE International Workshop on Computing in Civil Engineering. (pp. 581–588). Reston, VA: American Society of Civil Engineers.

37. Ray, FS In Morton AR (Ed.) (1984). Photographic optics. *Photography for the scientist, 2*, 104–124. Orlando, FL: Academic Press Inc.

38. Remondino, F, & Fraser, CS (2006). Digital camera calibration methods: considerations and comparisons. *International Archives of Photogrammetry, Remote Sensing and Spatial Information Sciences, 36*(Part 5), 266–272.

39. Rieke-Zapp, D, Tecklenburg, W, Peipe, J, Hastedt, H, Haig, C (2009). Evaluation of the geometric stability and the accuracy potential of digital cameras - comparing mechanical stabilisation versus parameterization. *ISPRS Journal of Photogrammetry and Remote Sensing, 64*(3), 248–258.

40. Rüther, H In Karara HM (Ed.) (1989). An overview of software in non-topographic photogrammetry. *Non-topographic photogrammetry, 10*, 129–145. Falls Church, VA: American Society for Photogrammetry and Remote Sensing.

41. Saidi, K, Cheok, G, Franaszek, M, Brown, C, Swerdlow, J, Lipman, R, Katz, I, Golparvar-Fard, M, Goodrum, P, Akula, M, Dadi, G, Ghadimi, B (2011). Development and use of the NIST intelligent and automated construction job site testbed, NIST TN – 1726 report.

42. Shen, X, Lu, M, Chen, W (2011). Tunnel-boring machine positioning during microtunneling operations through integrating automated data collection with real-time computing. *Journal of Construction Engineering Management, 137*(1), 72–85.

43. Shortis, MR, Robson, S, Beyer, HA (1998). Principal point behaviour and calibration parameter models for Kodak DCS cameras. *The Photogrammetric Record, 16*(92), 165–186.

44. Smith, M, & Park, D (2000). Absolute and exterior orientation using linear features. International Archives of Photogrammetry and Remote Sensing. XXXIII. 850–856.

45. Snavely, N (2010). Bundler: structure from motion (SfM) for unordered image collections. Code available at http://phototour.cs.washington.edu/bundler/
46. Tang, P, Huber, D, Akinci, B (2011). Characterization of laser scanners and algorithms for detecting flatness defects on concrete surfaces. *Journal of Computing in Civil Engineering, 25*(1), 31–42.
47. Tsai, RY (1987). A versatile camera calibration technique for high-accuracy 3D machine vision metrology using off-the-shelf TV cameras and lenses. *IEEE Transactions on Robotics and Automation, RA-3*(4), 323–344.
48. Viswanathan, M (2005). Measurement error and research design. Thousand Oaks, CA: Sage Publications.
49. Wackrow, R, & Chandler, JH (2008). A convergent image configuration for DEM extraction that minimises the systematic effects caused by an inaccurate lens model. *The Photogrammetric Record, 23*(121), 6–18.
50. Wackrow, R, & Chandler, JH (2011). Minimising systematic error surfaces in digital elevation models using oblique convergent imagery. *The Photogrammetric Record, 26*(133), 16–31.
51. Wegmann, H (2002). Image orientation by combined (A)AT with GPS and IMU. Proceeding of ISPRS Commission I Symposium. Denver, U.S.A. 10-15 November 2002
52. Zhang, Z (1999). Flexible camera calibration by viewing a plane from unknown orientations. *The Proceedings of the 7th IEEE International Conference on Computer Vision, 1*, 666–673.
53. Zhu, Z, & Brilakis, I (2009). Comparison of civil infrastructure optical-based spatial data acquisition techniques. *Journal of Computing in Civil Engineering, 23*(3), 170–177.

Chapter 11

Effects of the Tip Location on Single Piles Subjected to Surcharge and Axial Loads

Yaru Lv,[1] Xuanming Ding,[1] and Dubo Wang[2]

[1]Key Laboratory of Ministry of Education for Geomechanics and Embankment Engineering, Hohai University, 1 Xikang Road, Nanjing 210098, China

[2]North China University of Water Resources and Electric Power, 35 Beihuan Road, Zhengzhou 450045, China

ABSTRACT

When applying axial load on piles subjected to negative skin friction (NSF), the yielded NSF is gradually eliminated. The process is notably influenced by the tip location (Y) and still a lack of understanding.

This paper reports three-dimensional numerical simulations with tip locations Y=1.00 pile diameter (D), 0.25d, 0.00D, and -1.00D. It is found that, against expectations, the dragload and NSF are not proportionally related to the tip location. When maximum dragload (P_{max}) is eventually eliminated due to an axial load, there is also a negative crest of the skin friction, indicating that NSF still exists based on the criterion of the dragload reduction. The side resistance of the piles with Y=1.00D and 0.25D is almost fully mobilised, which is demonstrated by the increment of end resistance that greatly increases with the larger axial loads. However, the side resistance of the piles with Y=0.00D and -1.00D has a potential capacity to carry more loads with continued displacement since the increment of end resistance increases almost linearly with axial load. Therefore, when designing the pile foundation, the inclusion of the NSF should be governed by the amount of axial load to be resisted.

INTRODUCTION

Negative skin friction (NSF) due to the surrounding soil settling more than the pile has been reported by many researchers, who examined the NSF that occurs under the application of a surcharge load [1], withdrawal of the groundwater table [2], and/or soil consolidation [3]. The occurrence of NSF always induces an additional compressive force in pile shaft, which is called dragload. The caused dragload must induce additional pile settlement, which is called downdrag [4]. The location of zero skin friction is known as the neutral plane, where the pile-soil relative displacement is zero. Therefore, when designing pile foundations, it is unconservative when NSF is not considered, which can lead to serviceability problems and even eventual failures.

A number of tests have been carried out to investigate the behaviour of bored piles subjected to NSF [2,5–8]. Previous works have been conducted in the area of dragload on piles subjected to NSF due to a surcharge load or consolidating soil, but fewer studies have been conducted on an axially loaded single pile, let alone the dragload on an axially loaded pile [8–11]. In fact, there is lack of conditions in which piles are only subjected to surcharge loads. For example, although NSF is caused by the withdrawal of the groundwater table on piles carrying a tall building, the behaviour of NSF is heavily governed by the tall building (axial load). In recent years, several centrifuge tests have been conducted on piles subjected to surcharge and axial loads. These tests focus on the applied load conditions lack of considering the effects of the tip location [2, 3, 12]. As it is well-known, the tip location significantly affects the load transfer mechanism of piles subjected to NSF [13, 14]. It also governs the occurrence and elimination of the dragload and NSF, which is still a lock of understanding.

This paper studies the effects of the tip location on single piles subjected to a surcharge load and a series of axial loads, as shown in Figure 1. When subjected to only a surcharge load (Step 1), the surrounding soil settles more than the pile, leading to a dragload and NSF. The location of the neutral plane is governed by tip location, which is abbreviated as Y. When pile subjected to surcharge and axial loads (Step 2), the pile starts to settle and gradually exceeds the surrounding soil, leading to NSF and dragload elimination. Therefore, as an important factor, the tip location is considered to affect the degree of dragload elimination. However, how the tip location affects the occurrence or elimination of the dragload and NSF has not been studied and fully understood. Hence, three-dimensional numerical analyses are performed on piles with tip locations Y=1.00D,0.25D ,0.00D and -1.00D, where D is the cross-sectional diameter of the piles and -1.00D means the pile tip is located into the stiff sand layer with a depth of 1.00D. The numerical simulation with Y=1.00D is verified by reported measured data by Lam et al. [3].

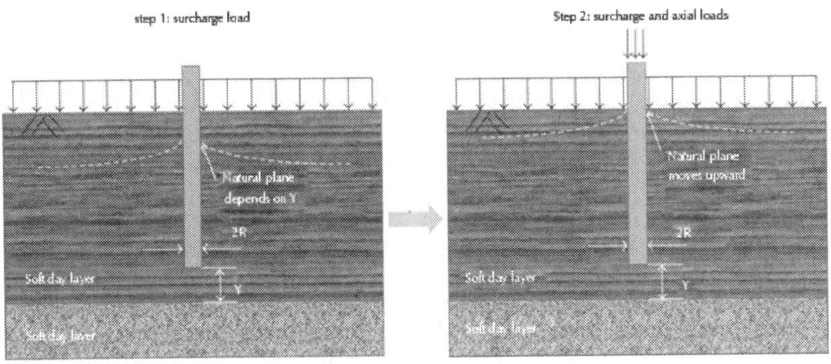

Figure 1: Graphical representations of a pile subjected to surcharge and axial loads.

NUMERICAL SIMULATIONS

Pile Geometry, Finite Element Mesh, and Boundary and Initial Conditions

Numerical analyses are carried out based on a reported centrifuge test [3] used for verification by means of a finite element programme called ABAQUS [15]. The three-dimensional element meshes and modelling scales are shown in Figure 2. Three-dimensional eight-nodded stress reduction elements are used to analyse the piles, and three-dimensional eight-nodded pore pressure elements are used to analyse the surrounding soil. Because of different tip locations, the number of elements and nodes are different. For the pile with Y=1.00D, there are 8,960 elements and 9,785 nodes for the soil and 896 elements and 1,189 nodes for the pile. Detailed information on the elements and nodes of the remaining three pile-soil systems is summarized in Table 1. To analyse the shear stress on the interface between the pile and its surrounding soil, fine meshes are generated for the soil near the pile.

Table 1: Summary of the elements and nodes in the numerical analyses

Numerical number	Tip location (Y)	Number of mesh		Numerical of nodes	
		Pile	Soil	Pile	Soil
1	1.00D	896	8,960	1,189	9,785
2	0.25D	928	5,632	1,230	6,373
3	0.00D	896	8,960	1,189	9,785
4	−1.00D	960	5,440	1,271	6,179

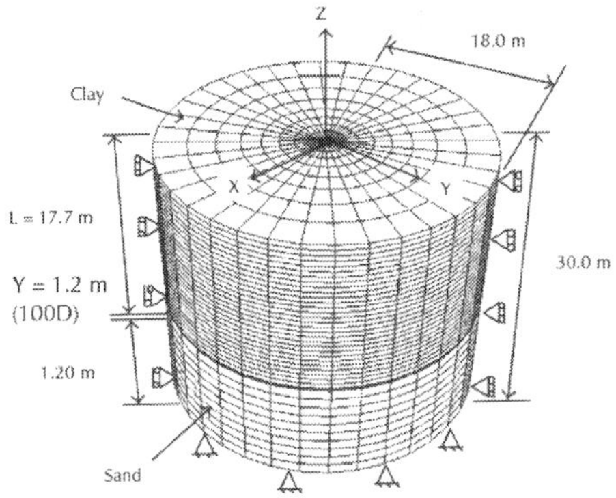

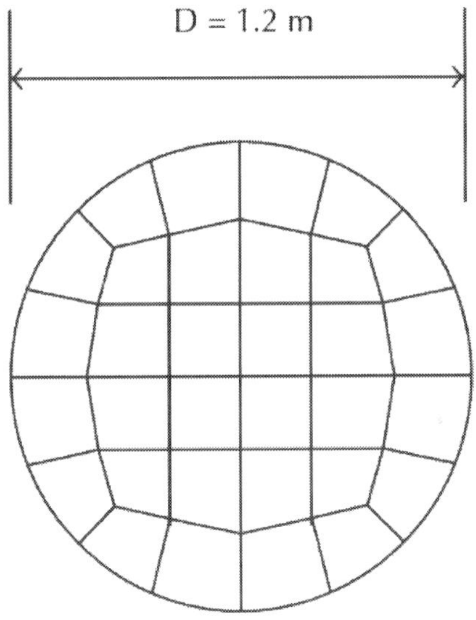

Figure 2: Finite element meshes adopted for: (a) the three-dimensional pile-soil system; (b) the cross-section.

The boundary conditions are as follows: the roller supports are specified on the lateral boundary elements and the pinned supports are specified on the bottom boundary elements. The water table is kept at the ground surface, which is a drainage boundary. 45 kPa pressures are applied on top of the clay layer to simulate the surcharge load. Coupled analysis is used in this paper. Hence, at the initial conditions, the initial effective stresses are set up using the buoyant unit weight.

Constitutive Models and Parameters

The constitutive models and parameters for each solid are shown in Table 2. The soil properties refer to the work of Lam et al. [14]. In numerical analyses, piles are simulated to be a linear elastic material, with a Young's modulus of (E_p=30GPa). The density and void ratio are 2.7 kg/m^3 and 0.18, respectively. The pile-soil slips are simulated using duplicated nodes to form an interface of zero thickness. The interface friction angle (δ) is estimated using the equation proposed by Randolph and Wroth [16] as follows:

$$\delta = \tan^{-1}\left[\frac{\sin\varphi\cos\varphi}{1+\sin^2\varphi}\right] \quad (1)$$

Table 2: Constitutive models and material parameters adopted for the numerical analyses

Solid	Pile	Bottom sand	Clay
Constitutive model	Linear-elastic	Mohr-Coulomb	Cam-clay
(kN/m³)	27	19.4	16.3
	0.18	0.30	0.35
Young's modulus, E (kPa)			N/A
M	N/A	N/A	0.98
	N/A	N/A	0.14
	N/A	N/A	0.012
(kPa)	N/A	N/A	64
	N/A	0.37	1.6
	N/A	29.7°	25°
	N/A	8.3°	0°
	N/A	0.50	0.58

Therefore, the interface friction coefficient of 0.32 is used in these analyses. The limiting relative shear displacement is taken as 0.005.

The soil is consisted by two layers, including 12 m sand underneath 18 m clay. The bottom sand is modeled as an elastoplastic material, with a Young's modulus of E_s before yielding. After yielding, the stress-strain relationships of the sand layer are described by a composite Mohr-Coulomb criterion. The sand is simulated as having a 70% relative density, corresponding to a saturated unit weight ($\gamma_{sat} = 19.4$ kN/m^3). Therefore, the value of p at the middle of the sand layer is approximately 100 kPa, corresponding to a maximum shear modulus of 103 MPa [17]. The soil stiffness significantly decreases with the shear strain, which is associated with foundation-related problems (typically approximately 0.1%) and is used to estimate the Young's modulus [18]. According to the empirical equation suggested by Ishibashi and Zhang [19], the shear modulus (G) associated with a shear strain of 0.1% is approximately 46 MPa. Therefore, the Poisson's ratio and Young's modulus of sand soil are taken as 0.3 and 120 MPa, respectively. The friction angle at the critical state (φ) is 29.7°. The dilation angle is 8.3° according to the empirical equation by Bolton [20]. The initial void ratio of 0.37 of the stiff sand is back-calculated by the water content [14].

For the top clay stratum, the failure criterion is described by the modified Cam-clay model, which was developed at Cambridge between 1960–1970 by Roscoe and Burland [21]. One of the basic characteristics of the model is non-recoverable material deformation (called plastic) after the nonlinear elastic response. Figure 3(a) illustrates the isotropic loading/unloading/reloading on the e-ln p plot. Experiments show that isotropic consolidation follows the isotropic loading line characterized by a material constant λ. Because of the elasto-plastic nature, the unloading path does not follow the consolidation loading line. Instead, the material follows an unloading line characterized by a material constant k. When the material is reloaded, it usually follows the same path as unloading until the mean normal stress p exceeds the maximum p in the loading history, and it then follows the load line again. The two material constants can be obtained by the following equations:

$$\begin{aligned} \Delta e &= -\lambda \Delta \ln p, \\ \Delta e^e &= -k \Delta \ln p, \end{aligned} \tag{2}$$

where e is the void ratio, p is the mean stress, and q is the deviatoric stress. Because a plastic volumetric strain increment can occur only when the current stress is on the yield locus, under isotropic loading, p equals p_0. Thus, the hardening parameter p_0 is a function of , ε_v^p, which is given by

$$d\varepsilon_v^p = \frac{\lambda - kdp_0}{1 + e_0 p_0} \tag{3}$$

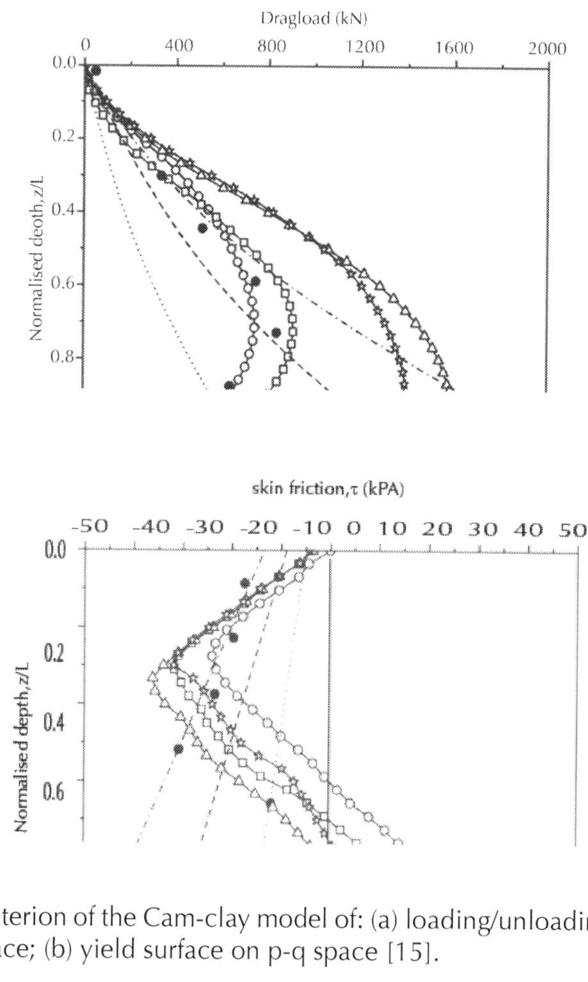

Figure 3: Criterion of the Cam-clay model of: (a) loading/unloading/reloading in e-ln p space; (b) yield surface on p-q space [15].

Figure 3(b) shows the yield surface of the modified Cam-clay model in p-q space. The yield function of the modified Cam-clay model can be written by

$$f = M^2p^2 - M^2p_0p + q^2 = 0, \tag{4}$$

where M is the slop of the critical state line and p_0 is the value of p at the intersection of the yield locus with the p axis, where q=0. The initial void ratio (e_0=1.6) is obtained by reported measurements [3]. The values of λ=0.14 and k=0.012 are specified through one-dimensional consolidation tests. Based on this test, the slop of the critical state line in p-q space (M) is given as 0.98 [3].

Modelling Procedures

As a setting for the simulation, the piles are installed wished-in-place (the piles are placed at the desired location without drilling a hole and applying the grouting material). The details of the simulation procedures are summarized as follows.

1. Create pile-soil numerical meshes and give each part the corresponding material properties.
2. Apply the boundary conditions.
3. Establish the initial stress conditions. The geostatic step is carried out with the lateral earth pressure ratio K_0=1-sinφ. After the geostatic step, the settlement of the ground is smaller than 10^{-5} m.
4. Apply a pressure of 45 kPa on the ground surface to simulate the surcharge load. Reveal the tip location effects on the occurrence of the dragload and NSF. This step lasts for 6.75 years for clay consolidation according to the centrifuge tests conducted by Lam et al. [14].
5. Apply the axial load (P) step by step with an increment of $P/P_{max} = 0.25$ (P_{max} is the maximum dragload which is obtained in step 4) to reveal the tip location effects on the elimination of the dragload and NSF, the variation of pile capacity, and the load transfer mechanism.

VERIFICATION OF THE NUMERICAL MODELLING

Tip location effects have been reported by Ng et al. [14], who mainly focus on revealing the differences between an end-bearing pile (Y=-1.00D) and a floating pile (Y=0.25D). The results show that the neutral plane of the floating pile is located at approximately z/H=0.75. For the end-bearing pile, the neutral plane is located near the bottom of the embedded pile length, causing NSF that almost acts on the whole pile. The result agrees with other studies [2, 7]. In the present analyses, to reveal the effects of the tip location, parametric studies are conducted on two floating piles with Y=1.00D and 0.25D and two end-bearing piles with Y=0.00D and -1.00D. The load condition changes from only a surcharge load to both surcharge and axial loads. The constitutive models and element meshes of the pile with are verified by means of the reported centrifuge test data [14].

Measured and Computed Dragload

Figure 4 shows the measured [14] and computed dragload of the piles with tip locations ,Y=1.00D , 0.25D,0.00D and -1.00D. To assist in the interpretation of the measured and computed dragload, the β -method is used to estimate the NSF [22–24], expressed by

$$f_s = \beta\sigma, \tag{5}$$

where f_s and σ are the NSF mobilised along the pile shaft and the effective overburden pressure, respectively. In soft clay, the value of β + is suggested as 0.25. Meyerhof [25] constructed a comprehensive summary on 20 sets of field data on values ranging from 0.18 to 0.35. Using (5), reference lines using three β values (0.1, 0.2, and 0.3) are calculated and plotted in the figure for comparison.

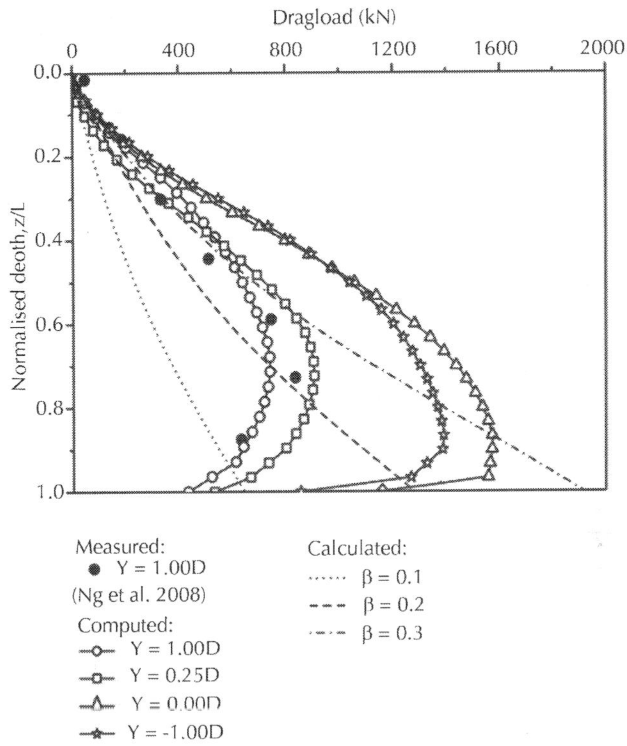

Figure 4: Measured and computed dragload distribution due to a surcharge load.

The measured and computed dragload increases at the upper depth and then decreases under the neutral plane. On the top of the neutral plane, the measured dragload [14] for the pile with Y=1.00D can be described by β =0.26, and the computed dragload can be estimated by β =0.30. It should be noted that the -method only gives a reasonable estimation of the dragload at the upper bound. It cannot take into account the reduction of the dragload at the lower bound.

The measured and computed maximum dragloads of piles with different tip locations are summarized in Table 3. Because the stiffness of the bottom sand is simulated substantially higher than that of the clay layer, the dragload on the floating pile increases with decreasing tip location. However, because the downdrag decreases with decreasing tip location, the dragload of the pile with Y=-1.00D is smaller than that of the pile with Y=0.00D. The measured and computed maximum dragload of the pile with Y=1.00D are 849 kN and 741 kN, respectively. The measured maximum dragload is 15% larger than the computed maximum dragload, indicating agreement between the numerical simulation and the centrifuge test. The computed maximum dragload of the piles with tip locations ,Y=0.25D ,Y=0.00D and Y=-1.00D are 907 kN, 1571 kN, and 1390 kN, respectively, which are 122%, 212%, and 188% that of the pile with Y=1.00D. In other words, the maximum dragload increases by 112% as the tip location ranges from Y=1.00D to Y=0.00D. Hence, the tip location should be as large as possible when floating piles are subjected to only a surcharge load. A large tip location can reduce the dragload and compression efficiently.

Table 3: Summary of the computed results

Results	Y=1.00D	Y=0.25D	Y=0.00D	Y=-1.00D
Maximum dragload (kN)	741 (849*)	907	1571	1390
(Maximum dragload/maximum dragload of pile with Y=1.00D) × 100%	100	122	212	188
Neutral plane, z/L	0.60 (0.73*)	0.71	0.82	0.75
Maximum soil settlement (mm)	55	75	75	55
Pile capacity (kN)	1080 (1728*)	1625	2822	3955
(Pile capacity/pile capacity of pile with Y=1.00D) × 100%	100	150	261	366

*Numbers in brackets denote the measured data.

Measured and Computed Skin Friction

The measured [14] and computed skin friction (τ) are illustrated in Figure 5. At the upper depth, the surrounding soil settles more than the pile, dragging the pile to settle with the surrounding soil. At the deeper depth, the pile tip inserts into the stiff sand layer, resulting in a larger displacement of the pile tip than its surrounding soil. Neutral planes exist at the interface of the downward and upward relative displacements. Hence, an important characteristic of the skin friction, the neutral plane, is yielded. The computed neutral plane is located at a depth of z/H=0.60 and 0.71 for the floating piles with Y=1.00D and 0.25D, respectively. Furthermore, the computed neutral planes are located at depths of z/H=0.82 and 0.75 for the end-bearing piles with Y=0.00D and -1.00D, respectively. This result indicates that the computed neutral plane shifts downward as the tip location decreases. However, when the pile tip is located in the stiff sand layer, the relative displacement between the pile and its surrounding soil is controlled, resulting in a smaller skin friction and a shallower neutral plane for the pile with Y=-1.00D compared to the pile with Y=0.00D. As a reference, the measured neutral plan is located at z/H=0.73 [14], indicating that the measured and computed results are found to be satisfactory.

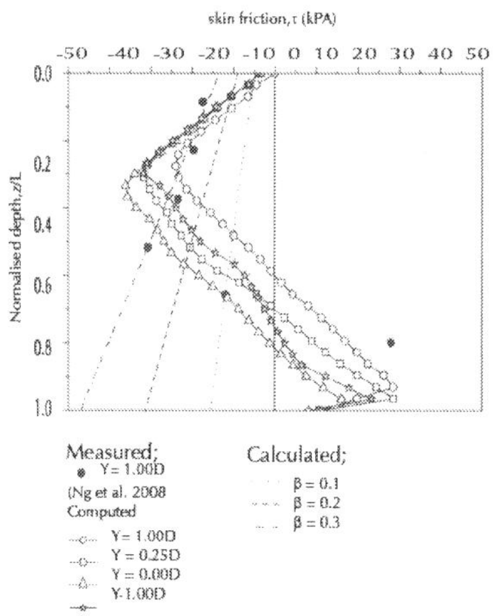

Figure 5: Measured and computed skin friction distribution due to a surcharge load.

Four piles have a similar trend that the computed NSF peaks at depth approximately z/H = 0.3. Meanwhile, the computed NSF peaks at depth approximately z/H = 0.5. Additionally, the measured positive skin friction (PSF) does not have a peak value, but the computed PSF peaks at about z/H = 0.9. This is because it is difficult to install the strain gauge continuously tomonitor the dragload and skin friction in the centrifuge test. Hence, there is lack of measured data near the pile tip, thus no peak value for the measured PSF. In this analysis, both the floating pile (Y = 1.00D and 0.25D) and end-bearing pile (Y = 0.00D and −1.00D) have PSF when subjected to the surcharge load. Each pile tip of four cases still settles more than its corresponding surrounding soil for the end-bearing piles. This phenomenon may be caused by two reasons. First, the relative stiffness between the clay layer and the sand layer is limited in this analysis. Second, the sliding effect of the pile-soil interface is used and affects the behaviour of skin friction, as discussed by Jeong et al. [13].

For reference, the calculated skin frictions by means of the β-method (β = 0.1, 0.2, and 0.3) are also shown in the same figure. As discussed in the dragload, the measured NSF can be described by β = 0.26 for the shallower depth. However, it cannot reveal the nonlinear variation of skin friction, especially for floating piles because the calculated skin friction varies linearly along the depth. Therefore, the β-method is more reasonable for estimating the NSF of endbearing piles because they have a limit PSF.

EFFECTS OF THE TIP LOCATION ON PILES SUBJECTED TO A SURCHARGE LOAD

As is well known, the occurrence of PSF and NSF is governed by the relative displacement between the pile and its surrounding soil. To show the effects of the tip location directly, the settlement of the surrounding soil in relation to the surface contour is plotted in Figures 6(a), 6(b), 6(c), and 6(d) for piles with tip locations Y = 1.00D, 0.25D, 0.00D, and −1.00D, respectively. When a surcharge load is applied to the surrounding soil, the displacement of the surrounding soil is larger than that of the pile at the upper depth. The negative relative displacement induces a downward-acting NSF. With increasing depth, most of the applied surcharge load is redistributed from the surrounding soil to the pile through the pile-soil interface.

The tip of the floating pile settles more than its surrounding soil because of the lower stiffness of the bottom soil (soft clay layer), inducing obvious upward-acting PSF. However, the tip of the end-bearing pile has limited settlement because of the higher stiffness of the bottom soil (stiff sand layer), leading to slight upward-acting PSF. Therefore, there is an obvious intersection between upward-acting PSF and downwardacting NSF, which is governed by tip location.

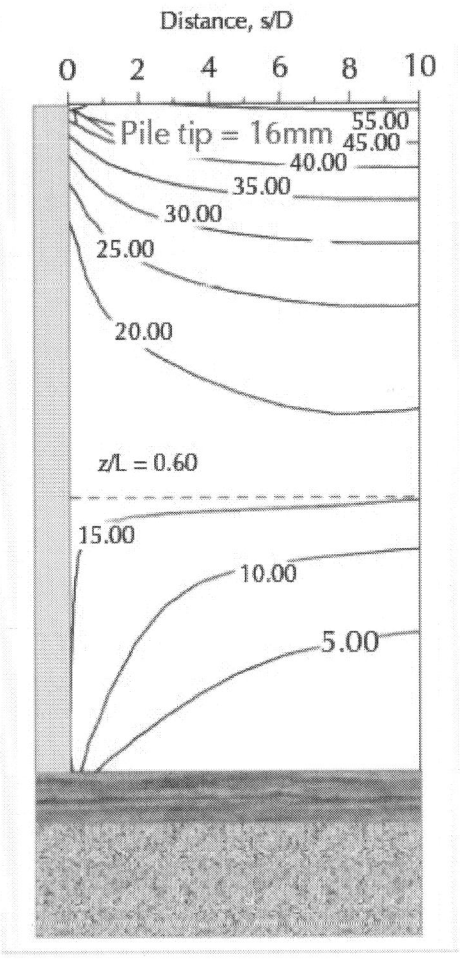

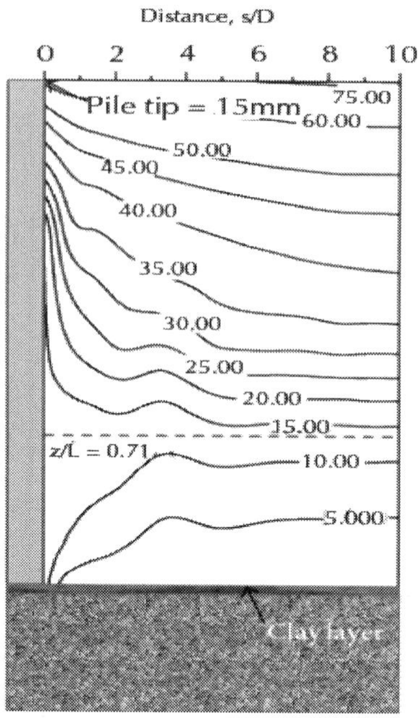
Distance, s/D
0
2
4
6
8
10
Pile tip = 15mm
75.00
60.00
50.00
45.00
40.00
35.00
30.00
25.00
20.00
15.00
z/L = 0.71
10.00
5.000
Clay layer

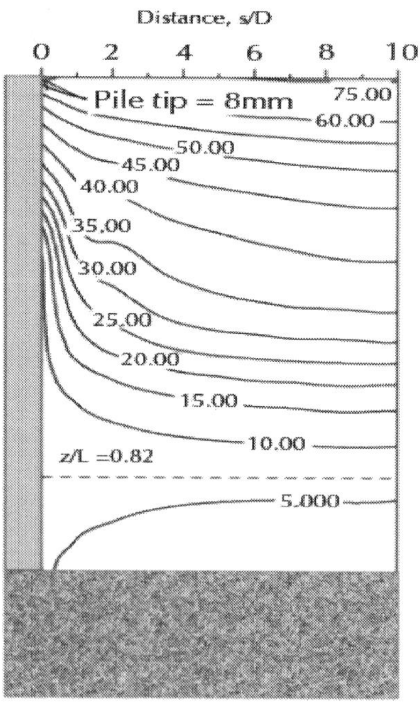
Distance, s/D
0
2
4
6
8
10
Pile tip = 8mm
75.00
60.00
50.00
45.00
40.00
35.00
30.00
25.00
20.00
15.00
10.00
z/L =0.82
5.000

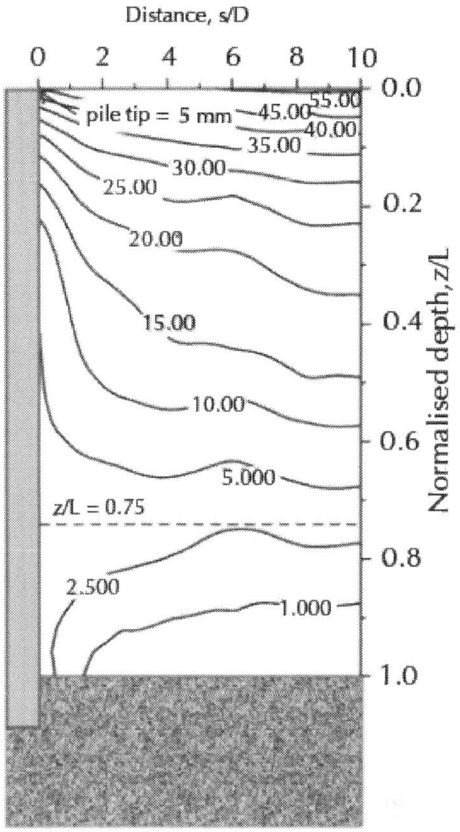

Figure 6: Settlement contour of the soil surrounding a pile with tip location: (a) $Y = 1.00D$, (b) $Y = 0.25D$, (c) $Y = 0.00D$, and (d) $Y = -1.00D$.

Against expectations, the settlement of the surrounding soil does not increase proportionallywith decreasing tip location, especially for end-bearing piles. When encountering only a surcharge load, both the surrounding soil and the pile with $Y = -1.00D$ settles less than the pile-soil system with $Y = 0.00D$, leading to a phenomenon in which the neutral plane of the pile with $Y = -1.00D$ is shallower than that of the pile with $Y = 0.00D$.

As summarized in Table 3, the computed maximum soil settlements are 55 mm, 75 mm, 75 mm, and 55mm for piles with tip locations of $Y = 1.00D$, $0.25D$, $0.00D$, and $1.00D$, respectively. On the other hand, the corresponding pile settle 16 mm, 15 mm, 8 mm, and 5 mm, respectively. It is concluded that the dragload and skin friction are heavily influenced by the relative pile-soil displacement. The larger the relative pile-soil displacement is, the larger the induced dragload and skin friction are for floating piles before yielding. From these contours, these contours, the locations of the neutral plane can be clearly illustrated. It is the zone surrounded by the downward settlement curve and the upward settlement curve.

A single pile influences the zone approximately 6D away from the pile [7]. Because of the drag effects induced by the pile, the settlement of the surrounding soil increases from the center to the far field. In this analysis, three piles with $Y = 1.00D$, $0.25D$, and $0.00D$ have a sphere of influence smaller than $6D$, as shown in the contours. However, the pile with $Y = -1.00D$ has a larger sphere of influence, approximately $8D$ far away from the pile, which may be caused by the embedded pile tip. For this case, the upper part of this pile is dragged by its surrounding soil, but the pile tip is difficult to insert into the stiff sand layer. Therefore, a large shear stress develops adjacent to the pile, resulting in vertical soil arching along the pile shaft. At the zone outside $8D$, the settlement curve is almost a horizontal line at a given depth.

EFFECTS OF THE TIP LOCATION ON PILES SUBJECTED TO SURCHARGE AND AXIAL LOADS

Measured and Computed Pile-Load Settlement

The settlement of a floating pile is more significant than that of an end-bearing pile [2, 6–8, 26]. The computed pile head settlements against an axial load are shown in Figure 7. The settlement is normalised by the pile diameter. The reported measured data (pile with $Y = 1.00D$) are taken into account and plotted for verification [12]. To reveal the process of dragload elimination, 20 steps of the axial load are applied to each pile with increments of 0.25 P_{max}; that is, $P/P_{max} = 0.25$, 0.50, 0.75, 1.00, 1.25, 1.50, 1.75, 2.00, 2.25, 2.50, 2.75, 3.00, 3.25, 3.50, 3.75, 4.00, 4.25, 4.50, 4.75, and 5.00.

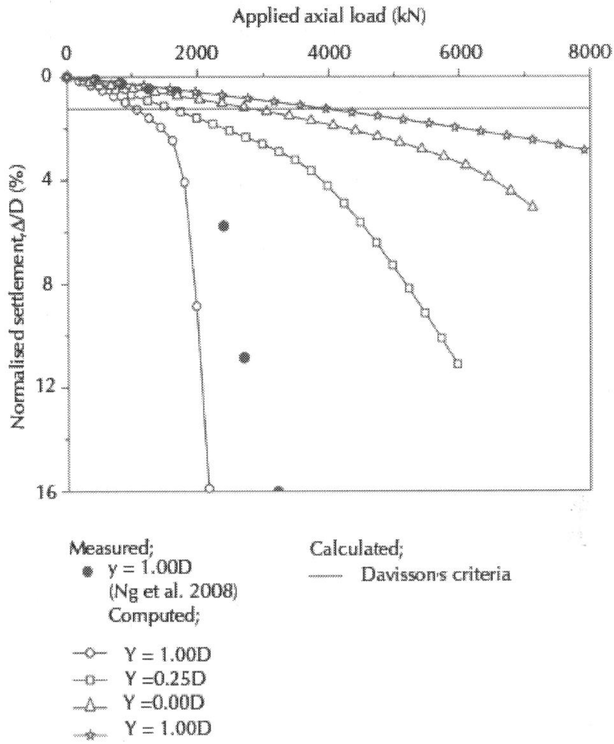

Figure 7: Measured and computed pile head settlement due to axial loads.

When designing deep foundations subjected to NSF, serviceability instead of the ultimate state usually becomes the governing factor. Because NSF is usually associated with a large settlement even with a small dragload (such as the pile with $Y = -1.00D$ being studied), excessive settlement could develop on shaft bearing piles with pile tips in compressive soils, which is true for the $Y = 1.00D$ case being studied. Davisson's criterion [27] gives the method for estimating the allowable loads for a pile. It can be written by

$$\Delta = \frac{PL}{AE_p} + \frac{D'}{120} + 4, \tag{6}$$

where L is the pile length (mm),Adenotes the cross-sectional area of the pile (mm2), E_p denotes Young's modulus for the material of the pile (kN/mm²), and D′ denotes the least lateral dimension of the pile (mm). Using this criterion, the normalised settlement of the piles being studied is = 1.24D without varying with tip location. Therefore, the allowable load, also the pile capacity in this study, is 1728 kN for the stated centrifuge test. For the computed load-settlement curve, the allowable load for the pile with Y = 1.00D is 1080 kN, which is approximately 63% that of the measured results. Both the measured and computed load-settlement curves have a similar trend. The pile-soil system yields at an axial load of approximately 1600 kN. When the axial load is smaller than 1600 kN, an almost linear variation of the settlements is induced, indicating the elastic compression that occurs in soils. When the axial load is larger than 1600 kN, both the measured and computed curves are far too steep to be taken as the serviceability condition. It is confirmed that the numerical simulations, including the constitutive models and element meshes, are reasonable.

As expected, the pile settlement decreases with the decreasing tip location. Two floating piles with tip locations Y = 1.00D and 0.25D have obvious elastic compression stages, with 1600kN and 3400kNbeing the yield axial loads, respectively. However, two end-bearing piles have no obvious yielding axial load until the last axial load step of, P/P_{max} = 5.00. Taking Davisson's criteria as the basis, the allowable loads of the piles with tip locations Y = 0.25D, 0.00D, and −1.00D are 1625 kN, 2822 kN, and 3955 kN, respectively, which are 150%, 261%, and 366% of that of the pile with Y = 1.00D, respectively. It is concluded that the tip location controls pile settlement and improve pile capacity efficiently. From Y = 1.00D to Y = −1.00D, the pile length (material waste) is increased by 14%, but the pile capacity can be improved by 266%. For this reason, the end-bearing pile is recommended when subjected to the axial loads.

Effects of the Tip Location on Dragload Reduction and the End Resistance Increment

As the load component changes from only a surcharge load to both surcharge and axial loads, the variations of the axial force for the piles with Y = 1.00D, 0.25D, 0.00D, and −1.00D are illustrated in Figures

8(a), 8(b), 8(c), and 8(d), respectively. As stated, the surcharge load yields the downward settlement of the surrounding soil, causing the piles to settle. Hence, the surrounding soil settles more than the pile, leading to a peak value of the dragload at the depth of the neutral plane. With different tip locations, the magnitudes of the dragload and the locations of the neutral plane are different.

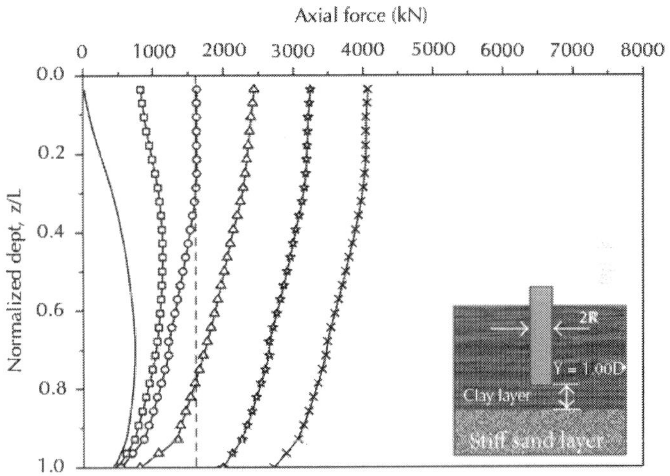

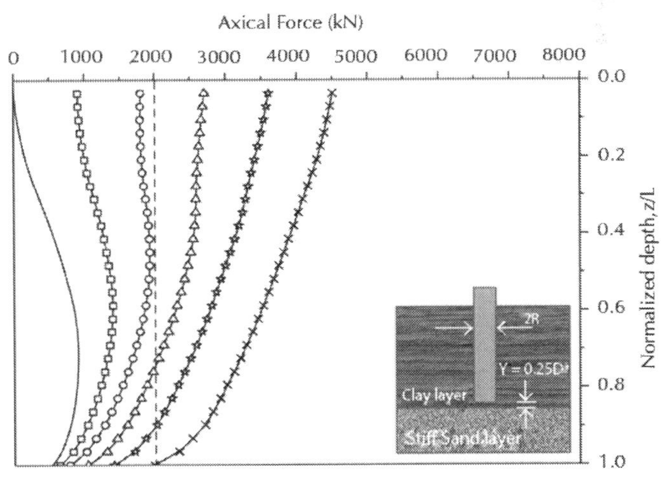

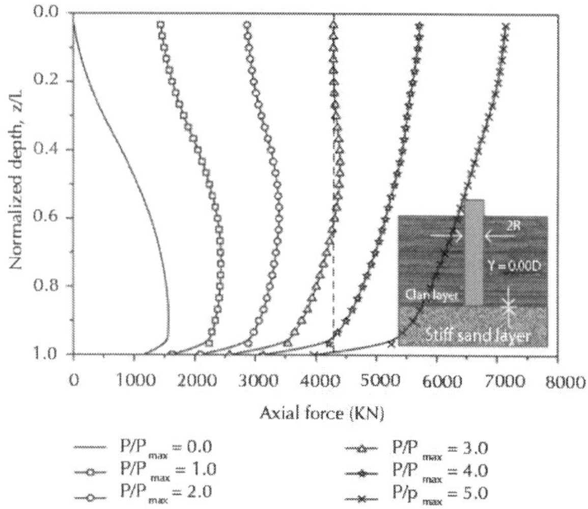

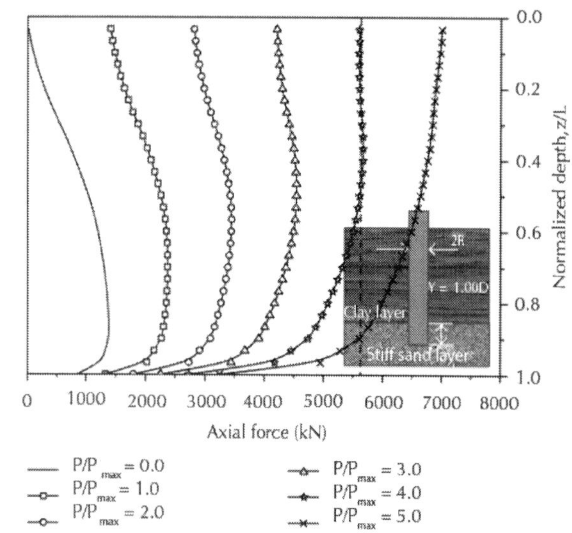

Figure 8: Computed variation of the axial force of a pile with tip location: (a) $Y = 1.00D$, (b) $Y = 0.25D$, (c) $Y = 0.00D$, and (d) $Y = -1.00D$.

As the axial load increases, a pile gradually settles faster than its surrounding soil. The position of the maximum dragload shifts upward, eliminating the dragload and NSF. Both the floating pile and the end-bearing pile have the same trend in that the maximum dragload eventually disappears, and is replaced by the maximum axial force at the pile head. However, it should be noted that the axial force along the pile shaft is not simply the sum of the dragload that is caused by the surcharge load and the axial force that is caused by the axial load. Under this load condition, the slope of the axial force and the proportion of the end resistance differ for different tip locations, which will be discussed later.

To describe the variation of the axial force from only a surcharge load to surcharge and axial loads, one coefficient, that is, the maximum dragload reduction, is defined. Jeong et al. [13] quantified the dragload reduction caused by the application of an axial load by using the following simple equation:

$$R = \frac{P_{max} - P_{max,axial}}{P_{max}} \times 100\%, \qquad (7)$$

where R is the rate of maximum dragload reduction; P_{max} is the maximum dragload before the application of an axial load; and $P_{max,axial}$ is the maximum dragload at a given axial load. The reduction of the maximum dragload for each pile is summarized in Table 4 and shown in Figure 9. It can be observed that the reduction of the maximum dragload varies almost linearly. The larger the tip location is, the more easily the dragload is completely eliminated. The maximum dragload reduction reaches 100%at approximately P/P_{max} = 2.0, 2.5, 3.5, and 4.5 for the piles with Y = 1.00D, 0.25D, 0.00D, and –1.00D, respectively. This result coincides with Jeong et al. [27], who reported that an axial load of approximately 1.25 to 3.25 P_{max} is required to eliminate dragload completely. For an end-bearing pile, especially a pile with the tip located in stiff bottom soil (as with the pile with Y = –1.00D being studied), although the magnitude of the dragload is limited due to the surcharge load, it is difficult to eliminate completely when subjected to axial loads.

Table 4: Variations of the dragload reduction and base stress increment with the axial load

Tip location (Y)	P/P_{max}																			
	0.25	0.50	0.75	1.00	1.25	1.50	1.75	2.00	2.25	2.50	2.75	3.00	3.25	3.50	3.75	4.00	4.25	4.50	4.75	5.00
R (%)																				
Y=1.00	16.2	30.8	45.0	57.1	68.7	79.4	89.2	100	100	100	100	100	100	100	100	100	100	100	100	100
Y=0.25	12.1	23.4	34.2	45.4	56.0	66.7	76.6	85.6	93.7	100	100	100	100	100	100	100	100	100	100	100
Y=0.00	11.7	22.5	31.0	36.9	44.7	52.1	59.3	66.3	73.4	80.2	86.8	92.9	98.9	100	100	100	100	100	100	100
Y=-1.00	8.8	16.8	24.2	31.0	37.3	43.2	48.8	54.3	59.9	65.3	70.8	76.0	81.2	86.1	90.7	95.2	99.7	100	100	100
I (%)																				
Y=1.00	2.8	5.5	8.0	10.8	12.6	14.9	18.3	21.2	23.9	37.2	56.6	85.5	125	217	310	357	403	449	494	525
Y=0.25	4.5	9.2	14.2	19.4	25.3	31.3	38.4	46.8	56.7	68.1	80.9	95.1	110	127	144	164	187	210	240	272
Y=1.00	10.6	22.5	32.6	40.4	50.3	60.3	70.2	80.3	90.6	100	111	122	134	145	157	170	184	201	221	242
Y=0.25	14.8	28.58	42.08	55.6	69.0	82.6	96.1	110	124	137	150	164	178	190	205	218	233	248	263	279

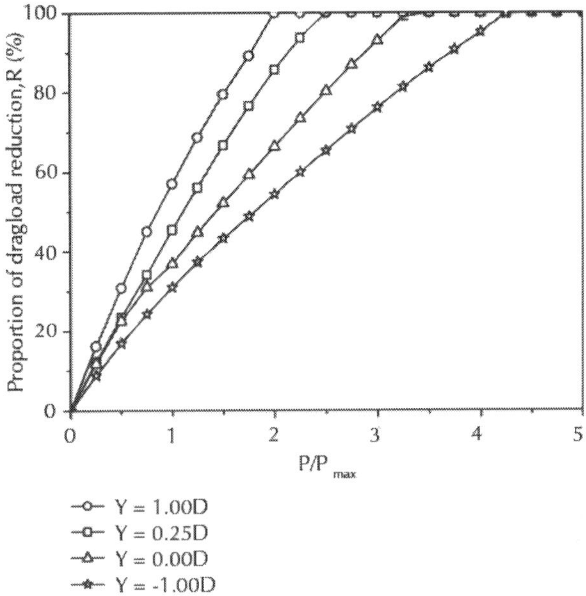

Figure 9: Computed maximum dragload reduction under surcharge and axial loads.

Furthermore, to describe the variation of the end resistance from only a surcharge load to both surcharge and axial loads, the other coefficient, that is, the end resistance increment, is defined as follows:

$$I = \frac{q_b - q_{b,axial}}{q_b} \times 100\%, \tag{8}$$

where I is the base stress increment, $q\ b$ is the dragload at the pile tip before the application of an axial load, and $q\ b$,axial is the axial force at the pile tip under a given axial load. The end resistance increment for each pile is summarized in Table 4 and shown in Figure 10. For the piles with Y = 1.00D and 0.25D, the increments of the base stress increase nonlinearly with the axial load and at an increasing rate. For the pile with Y = 1.00D, the end resistance heavily contributes to the pile capacity, which may be caused by the full mobilization of the side resistance at a larger axial load. For the piles with Y = 0.00D and −1.00D, the increments of the base stress increase almost linearly with the axial load. As stated, the maximum dragload is eliminated completely at P/P_{max} = 2.0, 2.5, 3.5, and 4.5 for the piles with $Y = 1.00D$, $0.25D$, $0.00D$, and $-1.00D$, respectively. Under the corresponding axial load, the increment of each end resistance is approximately 21%, 68%, 145%, and 248%, respectively. It is confirmed that the end resistance plays an important role for end-bearing piles before its dragload is eliminated completely.

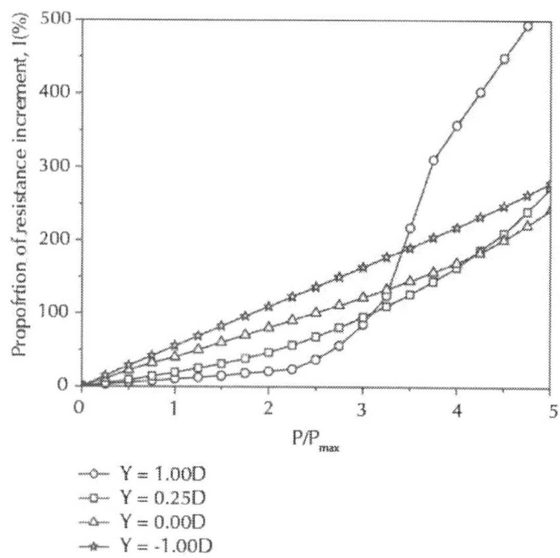

Figure 10: Computed end resistance increment under surcharge and axial loads.

From only a surcharge load to surcharge and axial loads, the variations of those two coefficients are governed by stress transfer mechanisms between each pile and its surrounding soil. Under a surcharge load, the surface load is carried by the soil and then transferred to the pile by means of NSF. Under both surcharge and axial loads, the load is carried by the pile and then transferred to the soil by means of the pile-soil interface. Therefore, both the maximum dragload reduction and the base stress increment result from the variation of the stress condition. Thus, when designing the pile foundation, the dragload should not be added to the loads from the structure and the capacity value should not be reduced by the dragload. The inclusion of NSF in the foundation design should be governed by the amount of axial load to be resisted. Exceeding the P/P_{max} value that the dragload is eventually eliminated, the additional compressive force should not be taken into account. Although the end-bearing pile has a higher pile capacity, the procedure of dragload elimination is difficult.

Effects of the Tip Location on Skin Friction

The computed variations of the skin friction are shown in Figures 11(a), 11(b), 11(c), and 11(d) for the piles with Y = 1.00D, 0.25D, 0.00D, and –1.00D, respectively. As stated in dragload, when subjected to a surcharge loadonly, the positionsof the neutral plane yield at z/L = 0.60, 0.71, 0.82, and 0.75, respectively. Whensubjected toboth surcharge and axial loads, the neutral plane moves upward. At a certain axial load, there are two neutral planes for the skin friction. This is caused by the relative pile-soil displacement. With increasing axial load, the pile first settles and then drags the surrounding soil to settle. This mechanism contradicts the stated mechanism of piles subjected to a surcharge load only. When the axial load is large enough, the neutral plane disappears. In this analysis, the neutral plane disappears at $P/$ Pmax = 2.0, 3.0, 4.0, and 5.0 for the piles with Y = 1.00D, 0.25D, 0.00D, and –1.00D, respectively. The load steps follow these values for the maximum dragload elimination.

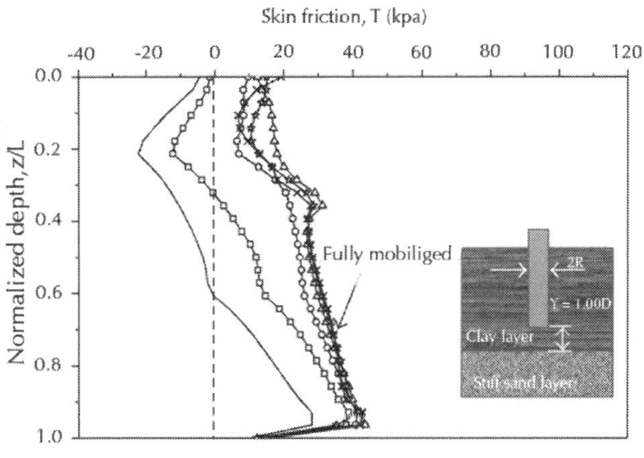

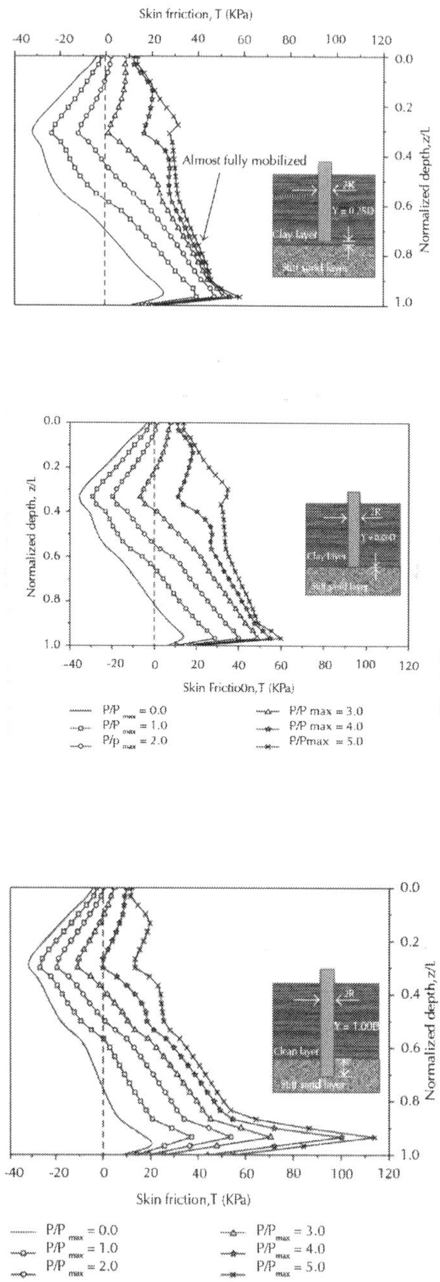

Figure 11: Computed variation of the skin friction of a pilewith tip location: (a)$Y = 1.00D$, (b)$Y = 0.25D$, (c)$Y = 0.00D$, and(d)$Y = -1.00D$.

For piles with different tip locations, the variation of the skin friction encounters different processes. For floating piles, the skin friction of the piles with $Y = 1.00D$ and $0.25D$ almost does not increase with increasing axial load after P/Pmax = 3.0 and 4.0, respectively. This indicates that the side resistance of eachpilehas almost been fully mobilised under the corresponding load step. This is the explanation of Figure 10, in which the increment of the end resistance greatly increases with the applied axial load after P/Pmax = 2.0 and 3.0 for the piles with $Y =$ $1.00D$ and $0.25D$, respectively. For end-bearing piles, the skin friction still proportionally increases with the axial load, which agrees with the increment of the end resistance in Figure 10. This phenomenon means that the side resistance has not been fully mobilised for end-bearing piles because of limited pile-soil relative displacement. In other words, the side resistance can be improved by sacrificing further settlement.

There is an unexpected occurrence that NSF due to the surcharge load is not fully eliminated when the maximum dragload disappears. Four piles have the same phenomenon that there is a negative crest of the skin friction, although the maximum dragload is fully eliminated. Therefore, when designing pile foundations, it should be noted that the NSF still exists based on the criterion of the maximum dragload elimination.

CONCLUSIONS

Three-dimensional numerical analyses are performed to investigate the effects of the tip location on piles subjected to surcharge and axial loads by considering the occurrence of the dragload and NSF, the elimination of the dragload and NSF, the variation of the pile capacity, and the load transfer mechanisms. The back-analysis of the pile with $Y = 1.00D$ is verified by the reported centrifuge test. The measured and computed results are satisfactory, indicating the reasonability of the constitutive models and element meshes. As the basis of those numerical analyses, several conclusions are drawn as follows.

1. As expected, the smaller the tip location is, the larger the dragload and skin friction are yielded for floating piles that the pile tip does not insert into the stiff funding soil. This is because the relative pile-soil displacement increases with decreasing tip location. Against expectations, the deeper the tip location is, the

smaller the dragload and skin friction are for end bearing piles that pile tip insert into the stiff funding soil. This effect is observed because the pile settlement is controlled by stiff funding sand. Hence, as the basis of this study, in order to reduce the dragload and the compression of pile shaft efficiently, the tip location of the floating pile should be as large as possible when meeting the requirement of the bearing capacity but the pile tip of the end-bearing pile should be located as deep into the bottom soil as possible.

2. The tip location plays an important role in controlling the pile settlement and improving pile capacity. From $Y = 1.00D$ to $Y = -1.00D$, the pile length (material waste) is increased by 14%, but the pile capacity is improved by 266%. For this reason, the end-bearing pile is recommended when subjected to axial loads without surcharge load or when the surcharge load has been fully eliminated.
3. When the load condition changes from a surcharge load to surcharge and axial loads, the developed dragload can be eliminated completely at $P/P_{max} = 2.0$, 2.5, 3.5, and 4.5 for the piles with $Y = 1.00D$, $0.25D$, $0.00D$, and $-1.00D$, respectively, and the corresponding NSF is fully eliminated at $P/P_{max} = 2.0$, 3.0, 4.0, and 5.0, respectively. There is a negative crest of the skin friction, although the maximum dragload is fully eliminated. In other words, it should be noted that when designing a pile foundation, NSF still exists based on the criterion of the maximum dragload elimination.
4. At $P/P_{max} = 2.0$ and 3.0, the skin friction of the floating piles with $Y = 1.00D$ and $0.25D$ almost does not increase with increasing axial load, respectively. This indicates that the side resistance of the floating piles is almost fully mobilised, as illustrated by the end resistance increment, which greatly increases with the axial load after $P/P_{max} = 2.0$ and 3.0, respectively. However, from$P/Pmax = 0.0$ to $P/P_{max} = 5.0$, the skin friction of the end-bearing pile still increases proportionally with the axial loads. This finding means that the side resistance of the end bearing pile can carry more axial load by sacrificing further settlement since the end resistance increment almost linearly increases with the applied axial load.

5. Therefore, when designing the pile foundations, the inclusion of NSF in the foundation design should be governed by the amount of applied axial load to be resisted and heavily influenced by the tip location.

When the axial load exceeds a certain value, which eventually eliminates NSF, NSF need not be taken into account.

ACKNOWLEDGMENTS

The authors thank Professor Ng and Dr. Lam, who supplied some valuable centrifuge test data for calibrating the numerical simulations of this paper. This work is supported by the Special Funds of Central Colleges Basic Scientific Research Operating Expenses under no.U1134207, the 111 Project under no. B13024, the Central Universities fund operating expenses no. 2013/B13020059, and support from Jiangsu province no. CXZZ12-0234.

REFERENCES

1. R. F. Shen, C. F. Leung, and Y. K. Chow, "Negative skin friction on end-bearing piles," in Proceedings of the 6th International Conference in Physical Modelling and Geotechnics (ICPMG '06), vol. 2, pp. 875–880, Hong Kong, China, 2006.
2. C. F. Leung, B. K. Liao, Y. K. Chow, R. F. Shen, and Y. C. Kog, "Behavior of pile subject to negative skin friction and axial load," Soils and Foundations, vol. 44, no. 6, pp. 17–26, 2004.
3. S. Y. Lam, C. W. W. Ng, C. F. Leung, and S. H. Chan, "Centrifuge and numerical modeling of axial load effects on piles in consolidating ground," Canadian Geotechnical Journal, vol. 46, no. 1, pp. 10–24, 2009.
4. B. H. Fellenius, "Pile terminology," 1999, http://www.geoforum.com/info/pileinfo/terminology.asp.
5. C. J. Lee, H. T. Chen, and W. H. Wang, "Negative skin friction on a pile due to excessive groundwater withdrawal," in Proceedings of the International Conference Centrifuge, vol. 98, pp. 513–518, Tokyo, Japan, 1998.

6. J. Thomas, M. Fahey, and R. J. Jewell, "Pile down-drag due to surface loading," in Proceedings of the International Conference Centrifuge, vol. 98, pp. 507–512, Tokyo, Japan, 1998.
7. C. J. Lee and C. Z. Chen, "Negative skin friction on ground piles," in Proceedings of the International Conference on Physical Modeling in Geotechnics, pp. 679–684, Newfoundland, Canada, 2002.
8. R. F. Shen, C. F. Leung, Y. K. Chow, Y. C. Kog, and B. K. Liao, "Negative skin friction on piles," inProceedings of the International Conference on Physical Modeling in Geotechnics, pp. 673–678, Newfoundland, Canada, 2002.
9. C. J. Lee and C. R. Chen, "Negative skin friction on piles due to lowering of groundwater table," Journal of the Southeast Asian Geotechnical Society, pp. 13–25, 2003.
10. Y. K. Wong, Numerical simulation of settlements associated with driven piles in Bangkok clay [MEng thesis], AIT, Bangkok, Thailand, 1991.
11. B. Indraratna, "Development of negative skin friction on driven piles in soft Bangkok clay: reply,"Canadian Geotechnical Journal, vol. 30, no. 5, pp. 887–888, 890–891, 1993.
12. S. Y. Lam, Effects of axial load, shielding and shape on negative skin friction on piles [M. Phil. thesis], Hong Kong University of Science and Technology, Hong Kong, China, 2006.
13. S. Jeong, J. Lee, and C. J. Lee, "Slip effect at the pile-soil interface on dragload," Computers and Geotechnics, vol. 31, no. 2, pp. 115–126, 2004.
14. C. W. W. Ng, H. G. Poulos, V. S. H. Chan, S. S. Y. Lam, and G. C. Y. Chan, "Effects of tip location and shielding on piles in consolidating ground," Journal of Geotechnical and Geoenvironmental Engineering, vol. 134, no. 9, pp. 1245–1260, 2008.
15. "ABAQUS standard user's manual," ver. 6.10, Hibbitt, Karlsson & Sorensen, Inc., Pawtucket, RI, USA, 2006.
16. M. F. Randolph and C. P. Wroth, "Application of the failure state in undrained simple shear to the shaft capacity of driven piles," Geotechnique, vol. 31, no. 1, pp. 143–157, 1981.

17. Z. Y. Cai, A comprehensive study of state-dependent dilatancy and its application in shear band formation analysis [Ph.D. thesis], Hong Kong University of Science and Technology, Hong Kong, China, 2001.
18. R. J. Mair, "Developments in geotechnical engineering research: application to tunnels and deep excavations," Proceedings of the Institution of Civil Engineers: Structures and Buildings, vol. 93, pp. 27–41, 1993.
19. I. Ishibashi and X. Zhang, "Unified dynamic shear moduli and damping ratios of sand and clay," Soils and Foundations, vol. 33, no. 1, pp. 182–191, 1993. View at Scopus
20. M. D. Bolton, "Strength and dilatancy of sands," Geotechnique, vol. 36, no. 1, pp. 65–78, 1986.
21. K. H. Roscoe and J. B. Burland, On the Generalized Stress-Strain Behaviour Wet Clay, Transportation Research Board, 1968.
22. I. J. Johanessen and L. Bjerrum, "Measurement of the compression of a steel pile to rock due to settlement of the surrounding clay," in Proceedings of the 6th International Conference on Soil Mechanics and Foundation Engineering (ICSMFE '65), vol. 2, pp. 261–264, Montreal, Canada, 1965.
23. J. Burland, "Shaft friction of piles in clay: a simple fundamental approach," Ground Engineering, vol. 6, no. 3, pp. 30–42, 1973.
24. H.-J. Kim and J. L. C. Mission, "Development of negative skin friction on single piles: uncoupled analysis based on nonlinear consolidation theory with finite strain and the load-transfer method,"Canadian Geotechnical Journal, vol. 48, no. 6, pp. 905–914, 2011.
25. G. G. Meyerhof, "Bearing capacity and settlement of pile foundations," Journal of Geotechnical and Geoenvironmental Engineering, vol. 102, no. 3, pp. 197–227, 1976.
26. S. H. Chan, Negative skin friction on piles in consolidating ground [M. Phil. thesis], Hong Kong University of Science and Technology, Hong Kong, China, 2006.
27. M. T. Davisson, "High capacity piles," in Proceedings of ASCE Lecture Series on Innovation in Foundation Construction, ASCE, New York, NY, USA, 1972.

Citations

CHAPTER 1

Congcong Li , Jie Wang , Lei Wang , Luanyun Hu and Peng Gong, Comparison of Classification Algorithms and Training Sample Sizes in Urban Land Classification with Landsat Thematic Mapper Imagery, doi:10.3390/rs6020964.

CHAPTER 2

Yi-Yuan Su, The Legal Structure of Taiwan's Wetland Conservation Act, doi:10.3390/su6129418.

CHAPTER 3

Chris Underwood, On the Design and Response of Domestic Ground-Source Heat Pumps in the UK, doi:10.3390/en7074532.

CHAPTER 4

Samuel R. Hamner, Ajay Seth, Katherine M. Steele, and Scott L. Delp, A rolling constraint reproduces ground reaction forces and moments in dynamic simulations of walking, running, and crouch gait, doi: 10.1016/j.jbiomech.2013.03.030.

CHAPTER 5

Li Tian, Wenming Wang, and Hui Qian, The Effect Analysis of Strain Rate on Power Transmission Tower-Line System under Seismic Excitation, doi: 10.1155/2014/314605

CHAPTER 6

L. Ran, X. W. Ye, G. Ming, and X. B. Dong, Structural Monitoring of Metro Infrastructure during Shield Tunneling Construction, doi: 10.1155/2014/784690.

CHAPTER 7

R. Doherty, B. Kulessa, A. S. Ferguson, M. J. Larkin, L. A. Kulakov and R. M. Kalin, A microbial fuel cell in contaminated ground delineated by electrical self-potential and normalized induced polarization data, doi: 10.1029/2009JG001131.

CHAPTER 8

Jason T. DeJong, Kenichi Soga, Steven A. Banwart, W. Richard Whalley,

Timothy R. Ginn,Douglas C. Nelson, Brina M. Mortensen, Brian C. Martinez, and Tammer Barkouki, Soil engineering in vivo: Harnessing Natural Biogeochemical Systems for Sustainable, Multi-functional Engineering Solutions, doi: 10.1098/rsif.2010.0270.

CHAPTER 9

Selma Kadioglu (2013). Transparent 2d/3d Half Bird's-Eye View of Ground Penetrating Radar Data Set in Archaeology and Cultural Heritage, Imaging and Radioanalytical Techniques in Interdisciplinary Research - Fundamentals and Cutting Edge Applications, Dr. Faycal Kharfi (Ed.), ISBN: 978-953-51-1033-0, InTech, DOI: 10.5772/54998.

CHAPTER 10

Dai et al.: Photogrammetric error sources and impacts on modeling and surveying in construction engineering applications. Visualization in Engineering 2014 2:2. doi:10.1186/2213-7459-2-2.

CHAPTER 11

Yaru Lv, Xuanming Ding, and Dubo Wang, "Effects of the Tip Location on Single Piles Subjected to Surcharge and Axial Loads," The Scientific World Journal, vol. 2013, Article ID 149706, 12 pages, 2013. doi:10.1155/2013/149706

Index

T

U

V

W